TURQUOISE

Mines, Minerals, and Wearable Art

Mark P. Block

4880 Lower Valley Road • Atglen, PA 19310

Library of Congress Control Number: 2017933324

Designed by RoS
Type set in Lato/Chaparral Pro

ISBN: 978-0-7643-5364-2
Printed in China

Other Schiffer Books by Mark P. Block:
Contemporary Marbles & Related Art Glass, ISBN 978-0-7643-1166-6
The Encyclopedia of Modern Marbles, Spheres, and Orbs, ISBN 978-0-7643-2294-5

Published by Schiffer Publishing, Ltd.
4880 Lower Valley Road
Atglen, PA 19310
Phone: (610) 593-1777; Fax: (610) 593-2002
E-mail: Info@schifferbooks.com
Web: www.schifferbooks.com

NATIVE AMERICANS believe that the earth is alive and that all things, no matter how small or seemingly inert, are precious. To the Native Americans, turquoise is life. Medicine men keep turquoise stones in their sacred bundles because the stones are thought to possess healing powers. The story goes that if you look down and see a crack in a piece of turquoise jewelry you are wearing, the Native Americans would say "the stone took it," meaning the stone took the blow you would have received. This book is for the rock hound and collector; miner and mineralogist; artist and craftsman who take pleasure in this beautiful and precious gemstone.

Legend has it that Native Americans danced and rejoiced when the rains came. Their tears of joy mixed with the rain before seeping into the Mother Earth, becoming "skystone," or turquoise.

Bronze Bust. Red Cloud with turquoise necklace. Location undetermined. *Courtesy of John Miller.*

Contents

Various single and multiple stone cuff bracelets. Bracelets include turquoise, coral, sugilite, and lapis. Silver setting. Signed John Renner. 1½"–3½" × 3". *Courtesy of the artist.*

New Mexico landscape. *Courtesy of John Renner.*

Display. JR's Southwestern Trading Post, New Hope, Pennsylvania. *Courtesy of John Renner. Photo by the author.*

An exciting display of various types, styles, and forms of turquoise jewelry from JR's Southwestern Trading Post on exhibit at the Palm Beach Jewelry and Antique Show, West Palm Beach, Florida. *Courtesy of John Renner.*

Foreword

WORKING with turquoise and being in the Indian turquoise jewelry business has been an amazing adventure, one that has led me to aid in this important project—documenting the various mines and varieties turquoise offers to the public.

I have been on a quest to search and find beautifully exotic older turquoise jewelry, pieces that exemplify the incredible diversity of Native American traditions and artists since my earliest days in the business. Equally as interesting for me has been showcasing, educating, and selling these beautiful pieces to new owners.

Over the years I have found great joy in cutting, polishing, and setting turquoise in sterling silver, and occasionally gold, to create objects of spirit and beauty. I created and staffed a lapidary and silversmith workshop in Albuquerque, New Mexico, and now spend most of my waking hours involved with turquoise in one way or another, either set in jewelry or in the rough. I buy it, sell it, or make it into jewelry. There has been no better education in identifying and appreciating the immense variety this beautiful gem has to offer.

As a jeweler and collector with an appreciation for high-quality turquoise, I am always on the lookout for the best available. Because I own a store, JR's Southwestern Trading Post in New Hope, Pennsylvania, and am often traveling to shows across the country, I don't have to limit my buying to just old or just new turquoise jewelry. I buy what I find personally interesting.

When people come up to my show booth at one of the indoor antique shows where I exhibit, they often ask where I got the amazing assortment of turquoise jewelry on display. I find them at estate sales, pawn shops out West, flea markets, the occasional wholesaler, and private sales; some pieces I design and make myself. Each piece is purposefully unique and has its own beauty.

Because many of the turquoise mines in America have closed or have limited production, the most affordable way to get good-quality American turquoise is to buy pre-1990s Native American jewelry. However, the supply is more limited every year. There are occasionally great bargains to be had, but usually you get what you pay for.

The dramatic increase in the price of silver was a bonanza for metal scrappers in the early 2000s, but a tragedy for lovers of turquoise and Indian silver jewelry because so many older jewelry pieces were scrapped. This also happened in the 1970s when the Hunt brothers tried to corner the silver market and drove the price of silver to $50 per ounce.

The contemporary turquoise jewelry market is dominated by imported turquoise, primarily from China. Good-quality Chinese turquoise is very beautiful and is often difficult to distinguish from some of the best American turquoise. This turquoise is increasingly being used for contemporary Native American jewelry. If you want to collect American turquoise or American turquoise jewelry, my best advice is to know who you are dealing with, and spend some time learning about turquoise itself. It is sometimes difficult even for experts to identify a particular piece. While there are identifying characteristics like color and matrix for individual mines, a mine can and often does produce some rocks that look closer to the characteristics of another mine. I have fifty pounds of natural Cerrillos turquoise with thick turquoise veins in large pieces. In a few rocks you can see on one side turquoise that is a beautiful sky-blue, and on the other, turquoise that is dark emerald-green. I would say seven out of ten turquoise stones are easily identified. The more turquoise you handle, the better you get at identifying the gem's origin. A reputable dealer or knowledgeable collector not only gives you an informed opinion but also helps further your education. They'll stand behind their opinion with a money-back guarantee. That dealer recognizes the grade and mine of origin and knows the fair market value.

Turquoise jewelry is not just for celebrities, although plenty of them wear and collect it. For thousands of years turquoise jewelry has been a fashion statement and was believed to have healing and protective powers. I always carry a piece of turquoise. You could call it habit or superstition, but turquoise has called my name throughout my entire life. A Navajo medicine man said that I didn't choose turquoise—turquoise chose me. He said my young life was a journey toward turquoise and my adult life has been a journey with turquoise. I am grateful to the core of my being for the great blessings that have come my way—the knowledge, experiences, and friendships—through a blue gem of beauty—turquoise. I wouldn't trade the adventure for anything in the world.

John Renner
New Hope, Pennsylvania

Lone Mountain turquoise. Bolo tie, 14k gold and silver setting. Raymond Yazzie (Navajo). C. 2012.

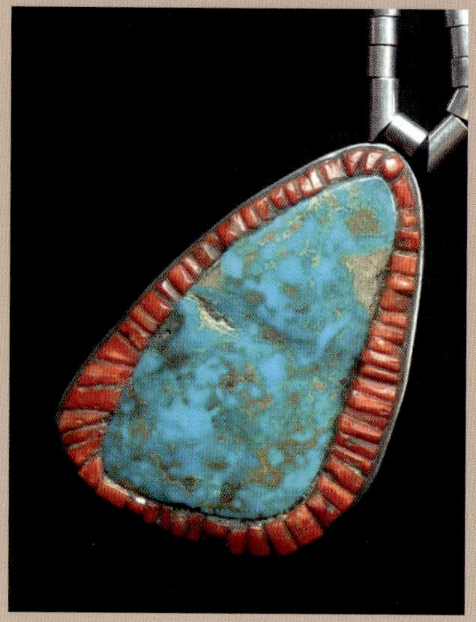

Blue Gem turquoise. Pendant necklace. Turquoise and coral. Silver setting. Unmarked. *Courtesy of John Miller.*

Unknown. Cuff bracelet. "Jellied" turquoise. Unknown origin. Multiple stones. Silver setting. Unmarked. *Courtesy of John Miller.*

Preface

WITHIN every collector lives a treasure hunter. The anticipation of the hunt, the excitement of the find, the passion in the soul—that's collecting in a nutshell.

Quality Native American turquoise in rough, cut and polished, and wearable form offers a unique collector/investor opportunity—as a collectible gem, an artistic statement, and a commodity. It is exactly this unique two-sided coin that attracts me to native turquoise jewelry. High-grade natural turquoise is a high-demand, limited-supply commodity and is currently seeing significant price appreciation. And fine Native American jewelry is a wearable art form comparable to that of the finest of Old European masters. Anyone with doubts need look no further than the work of this era's master, Navajo artist Raymond Yazzie; his creative genius and execution are second to none.

Successful collecting is rarely luck, but rather the result of hours of study and exposure to hundreds of quality examples, like the pieces displayed on the following pages. A discerning eye is the result of familiarity and attention to the details that expose that one piece that jumps out from the crowd.

I collect-invest based upon the following order of criteria: quality of the stone, quality of smithing workmanship, excellence of design, provenance of artist, and historical significance. When they all come together in one piece, it is exhilarating. All the long hours of study and searching and mistakes become sweetly worth it.

Every successful collector must learn to ask, "Is this a good find?" and more importantly, "Why?" There is no substitute for knowledge and developing a good eye. From cover to cover, *Turquoise: Mines, Minerals, & Wearable Art*, provides examples of high-grade turquoise set in exceptional-quality native jewelry, uniquely organized by turquoise mine origin. This will help collectors understand the nuances that separate fine turquoise native jewelry from tourist or cosmetic-fashion jewelry. By definition, a collector must be astute and settle for nothing less than the best. These pages highlight some of the best of the best.

John Miller
Anacortes, Washington
turquoisecollector.com

Acknowledgments

Sun Tracks
by Atoni (Choctaw)

The Track of the sun
across the Sky
leaves its shining message,
Illuminating,
Strengthening,
Warming,
us who are here,
showing us we are not alone,
we are yet alive!
And this fire . . .
Our fire . . .
Shall not die.

THIS poem provides an appropriate grounding for this book focusing on turquoise, its mines, and wearable art. While turquoise has been cut and set for centuries by peoples across the globe, it is particularly synonymous with the Native American culture of the American Southwest. This is evident in the blue and green, yellow and tan, cream and white colors in which turquoise is found, as well as the matrix colors of brown, black, or even red, offering a rich, illuminating, warming track of their own. As the words of the poem come together to tell a story, I have a sense of being transported to this region, one that continues to fascinate and captivate me. It has done so for more than four decades through the people I have met and known as much as by the landscape, art, history, and culture.

My first piece of turquoise came by mail more than forty years ago. It was inside a Chock Full 'O Nuts coffee can filled with agate, chalcedony, jasper, chrysocolla, and other minerals sent by a desert dweller, Marguerite Knickerbocker—first a long-distance pen pal and shortly thereafter one of the most giving friends I have ever known. She resided in the sleepy ghost town of Nelson, Nevada. No more than thirty people inhabited this "town" in the mid-1970s. At the dusty foot of the Eldorado Mountains, Nelson is nestled among the hills of southern Nevada, about forty miles south of Las Vegas. Nelson served as a hub of activity in the late-1800s, surrounded by some of the greatest and most prolific silver mines in the United States, including the famed Techatticup, which you can tour today, and Wall

Chinese, Kingman, and Sleeping Beauty turquoise. Various single and multiple stones, cluster cuff bracelets. Silver setting. Bracelets are offset by host rock with natural turquoise veins clearly visible. Pueblo and Navajo. 1¾"–2¼" × 3". *JR's Southwestern Trading Post.*

A look down Nevada State Route 165 toward Nelson in El Dorado Canyon, Eldorado Mountains. *Courtesy of the author.*

Nelson, Nevada, is about eight miles southeast of its junction with Route 95. Route 165 continues east five miles to a dead end at Nelsons Landing on the Colorado River, eighteen miles by water north of Cottonwood Cove on Lake Mojave. Nelson is about twenty-five miles from Boulder City and forty-three miles from Las Vegas by road. C. 1986. *Courtesy of the author.*

Nelson is a sparsely populated community consisting mainly of privately owned ranch houses, old mine shacks, mobile trailers, and a river and mining tour business housed in a former Texaco gas station north of the famed Techatticup Mine. *Courtesy of the author.*

Street Mines. Its history is legendary as the most violent town of the Old West, yet the inhabitants of the late twentieth century were some of the finest, most decent people you could ever meet, even if a road sign stated, "Don't Even Think About Stopping Here."

I owe a debt of gratitude and everlasting love to my late friend and "grandmother," Marguerite Knickerbocker. Her ability to withstand the harshness of winter and searing heat of summer were overshadowed by her loving and protective nature. She often told her friends, both locals and international performers and celebrities, that no one could find "such beautiful rocks" lying on the parched desert ground or in the hills and mountains of the canyon as I could. My eyes were always looking down; they still do whenever I make a trip from Las Vegas south and turn onto the winding road that leads down into the canyon, pay my respects at the boothill cemetery, and hike the canyon for a while, recalling days gone by. I delighted in showing her my treasures after my hikes. There's no doubt she would have loved this book.

Aldina Mang, Bill Hirst, Bob Hill, Carmen Methven, Bruno Liquori, Steve Liguori, and a host of Western "old-timers" enhanced my love for the American Southwest over the years. Sadly, nearly all have passed on and are at peace, yet every time I travel to the region I feel their presence and recall with warmth and quiet reflection the days and nights of a much sweeter time. A time when Las Vegas and the surrounding valley was less crowded, the neon truly vintage, the desert more open, and you could just about count the hotels and casinos on two hands. Each of these close friends imparted in me a sense of belonging, and often longing when we were apart. While I miss them, I am surrounded by their memories.

This book would not have been possible without the unmatched willingness and participation of John Renner, owner of JR's Southwestern Trading Post in New Hope, Pennsylvania. John has a unique perspective on both turquoise as a gemstone and its value as wearable art. Both from his shop in New Hope and during his travels across the United States, John was available and excited to work to ensure the first edition of this book contained a vast cross section of cut, polished, and beautifully set stones. This enables the reader to link the understanding of the gemstone with the notion that a single gem can have many variations in color and appearance. His expertise at identifying turquoise mines by simply viewing the stone made researching and writing the first edition a true labor of love. I would be remiss if I did not add a special thank you to Barbara and JoAnne as well for their aid in reaching the point of printing the first edition.

According to a Pew Research Center article, "Three Technology Revolutions" (2016), major technology revolutions have converged to change the way we gather and share information. This change is continuing to affect everything from social relationships to the way we learn, work, and take care of ourselves. It has also changed the way we collect. The speed of Internet connectivity has risen exponentially over the last decade, and with it people spend more time online. This revolution allows collectors to have a wealth of information at their fingertips. Through the Internet I gained access to professionals in the mineralogy field, which would have been much more difficult just a decade ago.

As I prepared to work on this updated and expanded edition I sought out collectors who could highlight areas not covered in the first edition. I was exceptionally lucky to have the assistance of John Miller of TurquoiseCollector.com. John has provided not only exceptional examples of turquoise from various mines, he has been a voice of professionalism on natural turquoise and turquoise mining. John's contribution to this volume adds another dimension of service to the collector, one that enhances the hobby and profession of turquoise collecting, dealing, mining, cutting, and setting.

Jim Olson, a ranch-raised cowboy and owner with his family of Western Trading Post in Casa Grande, Arizona, was generous in contributing fine examples of cut and polished turquoise wearable art. At Western Trading Post, founded in 1877, you can buy or sell, trade or pawn western items. Jim's family keeps the heritage of the West alive through their handling of Native American and other western collectibles and memorabilia.

I am blessed and fortunate to have a special relationship with Nancy Schiffer. With each book I have written, Nancy, and "my publisher," the late Peter Schiffer, extended their hands in friendship, knowledge, professionalism, and caring critique. Schiffer Publishing, with Pete Schiffer now at the helm, continues to broaden its willingness and faith in projects I propose.

My editor, Cheryl Weber, was always available and served as an important sounding board and guidepost to create a book that serves to benefit the written record and needs of the collector. The Schiffer family epitomizes the definition of friendship, and along with the entire staff at Schiffer Publishing, their encouragement and expertise has given me days of joy and the energy and fortitude to see the process of researching, writing, illustrating, and publishing to its bound conclusion.

A special thank you to the now retired Doug Congdon-Martin of Schiffer Publishing. When working

Bisbee turquoise. Deep-stamp cuff bracelets (top to bottom): Silver setting, Navajo, signed Delbert Gordon; Blue Gem turquoise and coral, Navajo, signed Ernest Roy Begay (ERB); Bisbee turquoise, Navajo, signed Jimmy a Yazzie; Bisbee turquoise, Navajo, signed Delbert Gordon. *Courtesy of John Miller.*

Cuff bracelets (inside). Deep stamp-work. Silver. *Courtesy of John Miller.*

on the first edition I would have hardly thought it possible for Doug to squeeze himself, along with his camera equipment, gems, jewelry, and even me all into one ten-by-ten room on-site in New Hope. Yet Doug never once complained. He produced some of the most beautiful images of turquoise jewelry ever to be published. My original editor, Jeff Snyder, also brought true professionalism to this book's first edition.

No one should write a book without the support and love of their family. I have been more than fortunate to have had both over my career as an author. I have been blessed with a loving wife and daughter who enjoy traversing the country to aid in photo shoots, research gathering, and appreciating the beauty of minerals, art glass, antiques, and collectibles. I've also been blessed with parents who instilled in my brothers and me the joy of collecting—the hunt really, from a very early age. From Brimfield, Massachusetts, to Renninger's in Pennsylvania, to Amana, Iowa, and points in between and beyond, memories for a lifetime were made in the back of a Ford Country Squire. Each book project has been a challenge to meet and a reward to read. It is all the more enjoyable to have them with me on life's journey, wherever it takes us.

Turquoise: Mines, Minerals and Wearable Art, first edition, was released with an accompanying Value Guide. At the time of publication this author believed that fair valuing guidance would be a valuable tool for the collector. However, prices fell during the Great Recession lasting from 2007 to 2009, and remained sluggish for years. Speculative price spikes of silver and gold, and the recent accelerated pricing for natural turquoise, the strongest since the popularity boom of the mid-1970s, has led this author to believe that a fixed-date value for the pieces illustrated would be valid only at the time of publication, and therefore would be a disservice to the collector. The best vehicles for current valuations are reputable dealers, gemologists, galleries, auction houses, specialty stores, and a myriad of highly regarded Internet sites.

Petite-point cut necklace, bracelet, ring. Silver setting. Zuni. Location undetermined. *Courtesy of John Miller.*

Introduction

Turquoise: Prized for Millennia

Look up at a clear, blue sky and you'll see quite vividly what has attracted people to the gemstone turquoise for millennia. Whether the sky is light blue in spring, deep blue in summer, or grayish blue in winter, each season astonishes us with its natural beauty, just like the myriad colors of this prized mineral. No other mineral is beloved for its robin's-egg-blue color, and it has come to be regarded as the "Tiffany of gems." Yet while this is the hue with which most people are familiar, turquoise exhibits many colors in the blue/green range.

This revised and expanded volume is intended to bring you closer to numerous types and styles of turquoise mined, cut, and set into jewelry—more than you may have ever seen before.

If you traversed the globe from ancient Persia through Mexico, and from North America to China, you would be dazzled by turquoise's color variations. Each region's mines produce different stones. In truth, the mines themselves can be distinctive, depending on the vein worked.

Would it surprise you to know that turquoise would produce little excitement without its distinctive range of colors? It is often described as a dull mineral. It is an opaque gemstone and not translucent like rubies, sapphires, or emeralds, yet it has held humankind's attention and respect as a prized gemstone for thousands of years. The very term "turquoise" has even taken on a meaning in culture as a recognized shade of blue all to itself. Adding to the history and importance of this gemstone, it is believed that turquoise was set into the breastplate worn by the high priest Aaron in biblical days. Its hold over people was enhanced when the treasures of the Egyptian pharaoh, Tutankhamen, were found to contain a significant amount of some of the most beautiful turquoise ever seen. The Egyptians favored this stone; its exceptional coloring is similar to that of the tropical sea and used to represent joy, cleanliness, and pleasure. Tutankhamen's infamous golden burial mask was inlaid with lapis lazuli, carnelian, and turquoise.

Turquoise continues to hold a place of honor among gemstones today. With more than 500 images in which to compare mined turquoise pieces with each

Morenci turquoise. Pendant. Silver setting. Navajo. 1½" × 1¾". *Courtesy of the author.*

Manassa turquoise. Pendant. Silver setting. Navajo. 1½" × 1½". *Courtesy of the author.*

other, this book should serve as an important reference in identifying the pieces in your own collection and those you would like to add. Until now, relatively little has been published on the turquoise mines themselves.

Let's turn the tables and see what awaits us as we focus on mines of exceptional reputation, such as Carico Lake and Cerrillos, Sleeping Beauty, Bisbee, and Kingman, as well as some that are more obscure, such as Ajax, Blue Jay, and Easter Blue. Each will give up their secret gemological treasures. Additionally, this book documents the importance of many lost or depleted mines to the annals of mineralogical history.

Mark Block
Trumbull, Connecticut

Kingman turquoise. 42 grams. *Courtesy of the author.*

Cuff bracelet. Early "pony" inscribed stamp. Silver setting. Location undetermined. *Courtesy of John Miller.*

Royston turquoise. Rough cut and polished. 33 grams. *Courtesy of the author.*

Sleeping Beauty turquoise, coral, shell, and black onyx. Rings. Multiple stones, inlay. Gold and silver settings. Zuni. 1"–1¼". *JR's Southwestern Trading Post.*

Tutankhamun's mask. Two layers of high-karat gold. 21" h, 15.5" w. × 19" d. C. 1323 BCE. Egyptian Museum, Cairo.

Miner's hut along Indian Creek near the Lander Blue claim. Bullion District, Lander County, Nevada. *Courtesy of John Miller.*

CHAPTER 1

THE MINERAL

TURQUOISE is classified as a hydrous phosphate, falling in the same category as apatite, mimetite, pyromorphite, and variscite, to name a few. Formed as water trickled through a host rock nearly thirty million years ago, deposits containing copper, aluminum, and zinc were left behind to form the beautiful gemstone turquoise.

The amount of copper in the minerals dictates the degree of blue that turquoise contains. This color can range from a pale powdery blue to dark blue, blue-green, matte green, and dark green.

However, when greater levels of aluminum are present the color tends to be more green or white.

If zinc occurs naturally in the mix the color becomes a yellow-green and the stone becomes even harder.

Properties

The best turquoise has a maximum hardness of just under 6 on the Moh's Scale. This makes the gem slightly harder than plain window glass. Rarely is turquoise found in well-defined crystals. Instead it is most often found in an aggregate of microcrystals. When the microcrystals become packed narrowly together, the turqoise is less porous, tougher, and polishes to a higher luster. This luster falls short of being vitreous or glassy. Instead it can best be described as waxy or suvitreous. Turquoise is nearly always opaque, though it can be seen as semi-translucent in thin sections. The color is as variable as its other properties. Colors range from nearly white to powder blue and sky blue, and can run from blue-green to yellow-green. The blue comes from idiochromatic

Blue Gem turquoise. Cut and polished. 8 grams. *Courtesy of the author.*

Natural copper specimen, Arizona, United States. 4" × 2¼".

Natural aluminum specimen, Russia. 2¼" × 1¾".

Natural zinc and sphalerite specimen, United States. 4" × 3¾".

copper, while the green may be the result of iron impurities that have replaced aluminum, or dehydration.

Turquoise forms best in an exceptionally dry climate, a determinant in the geography of prime sources. In these areas, rainfall infiltrates through surface soil and rock, dissolving small amounts of copper as it seeps down. When the water later evaporates, the copper combines with aluminum and phosphorus to deposit small amounts of turquoise along the walls of subsurface fractures. When larger amounts dissolve and the replacement becomes more complete, a bigger, more solid mass of turqoise forms, though if the replacement is less complete, the host rock will emerge as a matrix within the turquoise.

The refractive index of turquoise is approximately 1.61 or 1.62; and an absorption spectrum may be obtained with a hand-held spectroscope, revealing a line at 432 nanometers and a weak band at 460 nanometers. Under long-wave ultraviolet light, turquoise may occasionally fluoresce green, yellow, or bright blue; though turquoise is inert under shortwave ultraviolet and X-rays.

Turquoise is insoluble in all but heated hydrochloric acid. Its streak is a pale bluish white and its fracture is conchoidal, leaving a waxy luster. Despite its low to medium hardness, turquoise takes a clean, almost brilliant polish.

When other minerals appear in the host rock that turquoise was formed within, they are referred to as matrix. Matrix in turquoise rock can resemble a spider web or larger blotches. Spider web is matrix that is most often fairly evenly spaced or having an almost design-like appearance on the stone surface. It will often enhance the value and collectability of the cut and polished turquoise. A matrix of black is often caused by the introduction of pyrite, and yellow to brown matrix is caused by the introduction of iron oxide.

Easter Blue turquoise Pendant. Silver setting. Navajo, 2¼" × 1¾". *Courtesy of the author.*

Morenci turquoise and black onyx. Silver setting. Three-level custom designed ring. Zuni. 1½" × 1" × 1½". *Courtesy of the author.*

Occurrence and Association

From the romance language of French, the word turquoise is commonly translated as "Turkish." This is of interest not because of the mineral's color, but because turkeys and turquoise were both introduced to Europe through Turkey. The Mideast of old was a major producer of turquoise, and Persian turquoise is consistently sought out for its absolutely stunning, pure blue beauty. Just the name conjures up romantic images.

Dating the first discovery of turquoise is extremely difficult. Unlike Sutter's Mill in California, where the discovery of gold in the American West is well documented, it is not known when turquoise was first identified as a gem of appeal. What is generally recognized is that turquoise as a decorative gem appears to predate the Common Era by 5,000 years. For example, it is known that four bracelets belonging to Queen Zar and found on her mummified remains date to the second ruler of Egypt's first dynasty and were used for ornamentation over 5500 B.C.E. (Before the Common Era). As more and more discoveries were made in Egypt, turquoise beads were found in tombs. Later turquoise was inlaid with gold, the precious metal used extensively for kings and queens of the region to craft jewelry of the highest form by the finest craftsmen working for the rulers of their time.

Further, it has generally been acknowledged that the oldest known sources of turquoise were the Maghara Wadi mines in today's Sinai Peninsula. Thousands of laborers worked these mines, and the beautifully colored stones were meant for the pharaohs; in fact, these special mines would be worked for more than 2,000 years. Lost to time, the mines were rediscovered in the middle part of the nineteenth century and were mined occasionally until the early twentieth century. These mines have since been depleted and there is no active work occurring there today, though if you know your turquoise well enough it is possible to find a piece of this rare material, even in the twenty-first century.

Cerrillos turquoise. Cuff bracelet. Silver setting. 2" × 3". *JR's Southwestern Trading Post.*

As time passed, turquoise was found in many regions of the world. Archaeological history points to the Aztec kings having worn turquoise. It is unlikely that global trade with Egyptian pharoahs accounted for this. Not only is the time not correct, but also the trade route passages make this highly unlikely. The pre-Columbian Indians had beads and pendants made from turquoise. The color was so distinctive that the name turquoise has been used to describe any color that resembles it.

Turquoise first began to appear in North America around 200 B.C.E., where it was also used as ornament. The Southwest natives and Indian tribes of Mexico produced beautiful beads and necklaces from locally mined turquoise. This was a precursor to the explosion in turquoise jewelry that continues today.

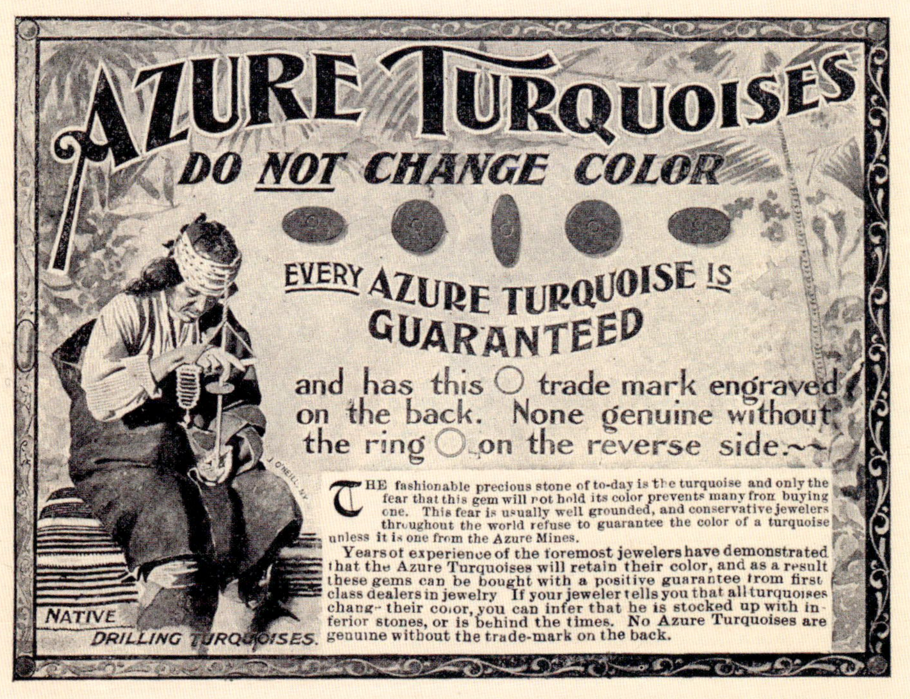

Azure Turquoise's magazine advertisement appeared in the early 1900s and stated, "Azure turquoises do not change color. Every Azure turquoise is guaranteed and has this O trademark engraved on the back. None genuine without the ring O on the reverse side." Image of *Native Drilling Turquoises.* 4½" × 3½". *Courtesy of the author.*

Ruins and other evidence indicates that turquoise was mined by both the Hohokam and Anazasi peoples. The material would have been found at the famous mines of Cerrillos and the Burro Mountains of today's New Mexico in the American Southwest. Other material has been identified as coming from Kingman (Arizona) and Morenci (Arizona), two of the most famous turquoise mine areas in North America. The Americas proved to be a fruitful trading ground, as Cerrillos turquoise has been found in Mexico, where the Aztecs used it, hundreds of miles from its mines.

Popular ideas about turquoise conjure images of gems mounted in silver by Native Americans—popularly known as Indian jewelry. It was the Spanish who first brought their knowledge of silversmithing to the Southwest, and combined with the lapidary knowledge of the Native Americans, an industry was born.

This industry remains quite separate from the religious and personal adornment pieces crafted by the Native Americans, and is a lucrative source of income for tribes including, Navajo, Zuni, and Hopi.

Jewelry making began in earnest in the late 1800s for the tourist trade in a region of the United States that was undergoing a vast transformation. Stages and trains bringing settlers from the East often stopped along the way. The trading posts that sprang up in these locations were very different from the basic posts of earlier days, as the newer posts serviced the newcomers with native jewelry. The natives often pawned their jewelry, and many of these antique pieces still appear in collections or at shows.

Vintage postcard illustrating turquoise drilling. Front reads: "Indian drilling turquoise, Pueblo of Zuni, Arizona." Back reads: "The Zuni Pueblo Indians are experts in making strand necklaces of shell and turquoise; some are silversmiths of merit and make many objects from American or Mexican coins such as rings, bracelets, belt buckles, etc., which find their way to tourists." 5¼" × 3½". *Courtesy of the author.*

Vintage postcard illustrating turquoise gem cutting. Front reads: "Largest gem-cutting house on the Pacific Coast. Southwest Turquoise Co., 113 N. Broadway. Los Angeles, Cal." 5½" × 3½". *Courtesy of the author.*

This jewelry is often referred to as "coin silver" turquoise, as it was crafted using the silver of melted coins. Many believe that this first occurred about 1880 when a white trader persuaded a Navajo craftsman to make turquoise and silver jewelry. Prior to this time, nearly all Native American jewelry was made of solid turquoise beads, inlaid mosaics, carvings, or other three-dimensional designs, and few, if any, pieces contained silver. Quite simply, the Native Americans had no access or ability to obtain coin silver. The vast silver mines of the West were being depleted about the same time, and silver bullion was worth vastly more than what would have been considered for use in Indian trinkets.

We don't often consider the importance turquoise played in Native American life. However, many objects were valued for their spiritual and practical purposes rather than as ornament. This is illustrated beautifully in the Native American belief that turquoise stone brings the spirits of the sky and sea to land, blessing both hunters and warriors. Although rather porous, a turquoise arrowhead was believed to ensure an accurate aim. The Navajo Indians believed that turquoise, cast with a prayer upon the waters, would cause rain to fall.

This is important to remember when evaluating the jewelry. It does not take a well-trained eye to differentiate the older, simpler pieces from the more contemporary, intricately crafted pieces, which often have artists' markings.

Candelaria and Kingman turquoise. Squash blossom necklace and matching earrings. Silver setting. Zuni, signed Ric Laselute. 18" length. *Courtesy of the author.*

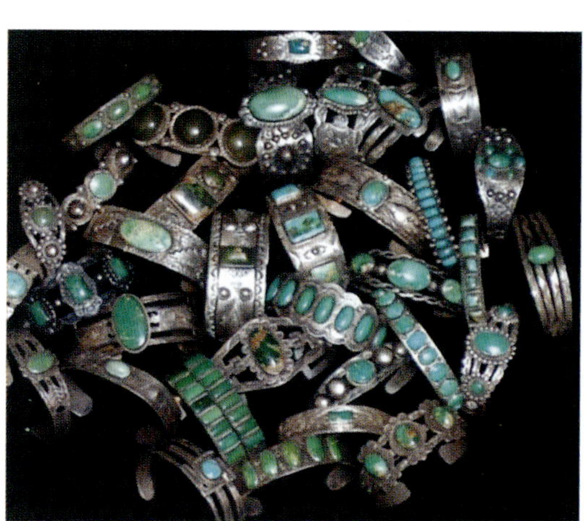

Assorted vintage 1930s turquoise bracelets. Silver setting. Includes Persian, Cerrillos, and Blue Gem turquoise. 1½"-2" × 3"-3½". *Photo courtesy of John Renner.*

Cerrillos turquoise. Cuff bracelet. Coin silver setting. C. 1910–1925. 1½" × 3". *Courtesy of the author.*

Tobacco canteen. Turquoise and silver. Unmarked. Location undetermined. *Courtesy of John Miller.*

Seed Pot. Turquoise and silver. Navajo, signed Roland Dixson. Location undetermined. *Courtesy of John Miller.*

Seed Pot. Turquoise and silver. Navajo, signed Roland Dixson. Location undetermined. *Courtesy of John Miller.*

Seed Pot. Inlay Story Teller. Turquoise and silver. Blue Ribbon (Santa Fe Indian Market). Navajo, signed Sunshine Reeves. Location undetermined. *Courtesy of John Miller.*

Octagon box. Turquoise and silver. Navajo, signed Sunshine Reeves. *Courtesy of John Miller.*

Royal Web turquoise. Bracelet and ring. Joe Tanner mine. Silver setting. Navajo, signed Robert Sorrell. *Courtesy of John Miller.*

Seed Pot. Turquoise and silver. Navajo, signed Roland Dixson. Location undetermined. *Courtesy of John Miller.*

Robert Sorrell hallmark. *Courtesy of John Miller.*

CHAPTER 2

EVALUATING TURQUOISE

WHETHER turquoise exhibits the solid blue of Bisbee (Arizona), the intense greens of Cerrillos (New Mexico), or the pale blue/greens of Number 8 (Arizona), each variation has its own innate beauty. It would be unfair to compare the finest Lander's (Nevada) turquoise with that mined from China, as each turquoise mine is unique and has produced gems of both historical and decorative value. Turquoise was so highly valued that the crown given by Napoleon I to Empress Marie Louise had its seventy-nine emeralds removed and replaced with Persian turquoise cabochon stones.

Grades

When grading turquoise, consider the stone's hardness or density. Turquoise ranges from 2.0 to nearly 6.0 on the Mohs scale; its gravity also varies but is usually 2.8, like quartz. The harder the stone, the better the cut and polished gem. The stone's luster should come both from the material and its polish. When considering color, be aware of the hue and intensity, whether the color is blue, green, white, or tan. Likewise, the matrix, or pattern, should be pleasing or unusual and accentuate the stone. Lastly, rarity can be immensely important The more difficult it is to find a particular mined turquoise, the higher the value.

Turquoise as a gemstone is classified in three ways: natural, stabilized, and treated. While values fluctuate wildly among the three classes, the ability to identify a piece's grade can prevent a collector from paying more than it is worth.

Natural turquoise. The most judicious collectors strongly prefer natural turquoise—that is, the gemstone is hard (Mohs scale 5.0–5.6) and has not been altered or enhanced. By definition, gem-grade stone does not need alteration. Turquoise that contains a high level of silica has a dense, non-porous surface that makes it impenetrable to treatment. For the most discerning collector there is no other form of turquoise that is considered collectable. Natural turquoise means no artificial enhancement of any kind has been done or taken to alter the stone.

This is not to say it is not polished—it is. The mines of the American Southwest, and more recently China, have produced outstanding gems that have been cut and polished, designed, and set in jewelry of the finest quality. Many pieces are enhanced, however, and it is important that this is clearly identified. As with any gemstone, deceptive representations take place. The collector should be well aware of the presence of inexpensive Asian knockoffs.

Treated turquoise. This term refers to a process of hardening the stone and deepening the color. Light-colored turquoise, the most prevalent, is softer and more porous—often nearly chalk-like—than higher-quality varieties. This can vary from mine to mine. Turquoise that is less dense will remain porous and have a tendency to change color as a result of wear and exposure to light. Thousands of years ago, animal fat was used to harden turquoise and deepen its color. Today this is accomplished by submerging it in a stabilizing compound, usually an epoxy-type resin. This process is not to be confused with heat-treated, synthetic, or composite stones.

Natural crystalline turquoise specimen. Bishop Mine, Lynch Station, Virginia. 2½" × 1¾". *Courtesy of the author.*

Stabilized turquoise. Often you will hear the terms "stabilized" and "treated" used interchangeably. While the process may be slightly different, the effect is the same—to harden or stabilize soft, porous turquoise. This process involves filling porous areas of the material with artificial enhancers like epoxy or another polystyrene chemical, which creates a shinier surface. What occurs is a stabilization or freezing of the color so changes will not occur later and the stone can then be cut and used. This is the most common and often necessary means by which to make turquoise usable as a gemstone. More than eighty percent of all turquoise has been stabilized before being cut and polished.

Stabilizing turquoise may also involve backing the stone with epoxy, what is called a doublet. Because nearly all turquoise is mined from veins, much of it is extremely thin, making it impossible to cut for use in jewelry. The backing enables the stone to be protected from cracking, chipping, or breaking when it is cut and set. This is a perfectly acceptable way of ensuring that a rare or extremely beautiful piece of turquoise does not end up on the scrap heap. Many lapidary and jewelry artists do this to create works of accomplished beauty from material that would otherwise be unusable. Sadly, most of the turquoise mined over the last hundred years or so would be relatively worthless if it weren't for the ability to stabilize it, unlike the fine gems of ancient Persia, the mecca of turquoise mining.

Enhancing is a newer term applied to turquoise that has been hardened and color-enhanced through an electrical process. This alters its chemical structure and is referred to as the Zachary or Foutz process of vaporized quartz alteration. It alters the potassium level, thereby making the turquoise no longer natural and thus no longer rare. The true collector of natural turquoise may only add this particular form if it is used in a specific artistic style or piece, like Zuni figurative inlay.

Blue Gem turquoise. Stabilized. 80 grams total. *Courtesy of the author.*

Kingman turquoise. Color-treated. 15 grams. *Courtesy of the author.*

Other Alterations

Dyed turquoise. Some lower-quality turquoise is infused with artificial elements that keep it from regressing to its natural brittle state and enhances the color. The process is similar to stabilizing turquoise, only in this instance the hardening agent is mixed with dye—typically a robin's egg blue color. After pressure is applied to infuse the material with the dyeing agent, the stone has a plastic appearance. This is the best way to identify turquoise that has been chemically altered. It also looks like an imitation because of its blue color and highly polished appearance. Another easy way to distinguish this low-grade turquoise is to take a knife blade or open paper clip and scratch the material. An easy scratch is a good indication the turquoise has been enhanced or treated. Higher quality turquoise cannot be scratched easily with a knife even though turquoise falls mid-range on the Moh's scale of hardness. The value of this chemically treated turquoise is considerably less than natural, treated, or stabilized material.

Reconstituted turquoise. Cuff bracelet. Turquoise, coral, shell, and black onyx, multiple stones. Silver setting. Navajo. 2¼" × 3". *JR's Southwestern Trading Post.*

Block turquoise. Earrings. Multiple stones. Navajo. 3¼" × 1½". *JR's Southwestern Trading Post.*

Reconstituted turquoise. This is a way to save all the scraps of mined or ground material. The turquoise dust, chips, and bits are mixed with plastic resins and compressed into a solid form. This man-made material is acceptable for use in jewelry and carvings, but again, it is important to know that the material is not in its natural state. The color of reconstituted or block turquoise is often quite beautiful itself.

Block turquoise contains no turquoise at all. Like reconstituted turquoise, it is man-made with plastics and dyes. Other minerals can be added to this block form, including pyrite, to create an interesting matrix when cut and polished. This material is most often used in inlay work and beads. Block turquoise is commonly mistaken for reconstituted turquoise.

Imitation turquoise is just that. For years, plastic has been a favorite for use in costume jewelry and beadwork. It can be formed to imitate natural turquoise, though in most cases, its luster is much waxier. The most often used material has, in fact, been another mineral—howlite. This mineral has the look of spider web turquoise when dyed blue. It is also extremely porous and accepts dye easily. The novice can easily confuse dyed howlite with natural or stabilized turquoise.

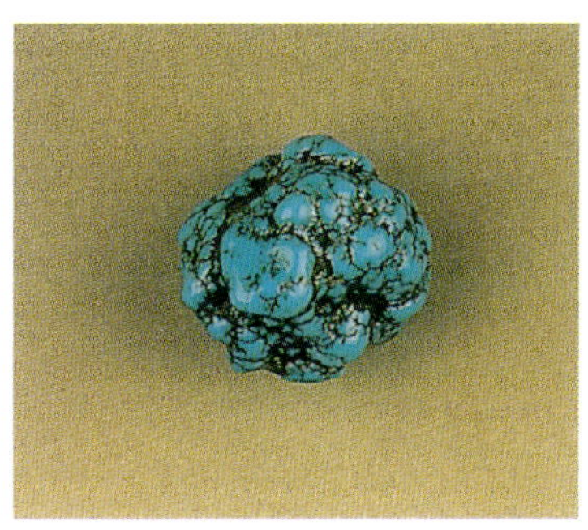

Howlite. Dyed. 8 grams. *Courtesy of the author.*

Other minerals used to imitate natural turquoise include chrysocolla, amazonite, variscite, and chalcosiderite. Additionally, be aware that glass imitations also appear on the market and are distinguishable by their high luster.

Reconstituted, block, and imitation turquoise is not generally collectable because it is man-made. It is important to pay particular attention to the consistency of color and the matrix patterns. Natural turquoise is not uniform—this is a sign of mass production. In short, no two turquoise stones will be exactly alike.

Value

Valuing is done by considering a number of factors. Primarily, is the turquoise itself natural and rare? Does it come from a discontinued mine source or one that produced little material? Is the turquoise of exceptional quality, meeting all of the criteria for natural color, cut and polish, and setting? These are just some of the questions to ask when searching for turquoise jewelry and rough or cut and polished stones.

More stable than monetary currency, fine gems and jewelry have been an important commodity for centuries. Few would argue that, in general, gems have proven to be a good investment. In turquoise, it is not likely that you will find many bargains. Its availability is limited, and certainly high-quality natural turquoise is even more difficult to find, whether through a wholesaler or retailer.

Part of the allure of turquoise is knowing that it is the oldest gemstone mined in the United States. The cultural connection to the American Southwest is extremely strong, and many can identify with the romanticism of earlier days involving the American Indian and West.

Of course for turquoise to have value it must have visual appeal. The character of a piece must elicit a response, from the stonecutter and polisher to the wearer or collector. This occurs with turquoise from small mines—for example, the mere bucketfuls that came from Landers (Nevada) and the desirable Blue Gem (Nevada) mine. This is in contrast to the stunning material that was mined for decades from Bisbee (Arizona), Kingman (Arizona), and Sleeping Beauty (Arizona). Each mine produced turquoise that varied in quality, yet was quite distinctive and sought after by Native American and other jewelers.

The color of a turquoise stone is the essential component in assessing the value of a piece. A matrix of spider web, pyrite, or other inclusions can enhance or detract from the piece, though this is subjective. The size, cut and polished shape, and quality are important factors in attributing value to any piece beyond its color. Also, the stone's setting can enhance the value dramatically. Has it been singly set or set with other gemstones? Is the piece part of a set—for example, a pendant and matching earrings? Is the piece antique or contemporary? And if contemporary, is the artisan a well-known and respected craftsman? Has the piece been exhibited in any shows or won any awards? And finally, consider the piece's carat weight, whether it is an individual stone or the combined weight of several stones in a piece of jewelry.

Indian Fair Blue Ribbon. Heard Museum. Indian Art Juried Competition. Belt buckle from Story Belt titled, "She begins to weave, She becomes Spider Woman." 2013 Heard Blue Ribbon Award (Belt Division). Buckle depicts Navajo Legend of Spider Woman Descending Spider Rock (Canyon de Chelly) to Teach the Dine to Weave. In this scene, Spider Woman is returning up Spider Rock, having taught the Dine the art of weaving. *Courtesy of John Miller.*

The entire north side of the downtown Plaza in Santa Fe, New Mexico, is taken up by the Palace of the Governors. Built in 1610, it's the oldest continually occupied public building in the United States. Its front adobe façade is completely shaded, and in this "portal" the Native American Vendors Program has been operating for more than six decades. A daily lottery ensures a rotating selection of artisans from the various pueblos throughout New Mexico. *Courtesy of the author.*

Palace of the Governors

Turquoise, like other gemstones, is measured and sold by carat weight. A unit of measure to remember is five carats per gram. In today's market, the price of turquoise can range anywhere from fifty cents per carat to hundreds of dollars per carat. The value will vary with economic conditions, the vagaries of the market, and the quality and provenance of the collection. Though like any collection, the most important tip is to collect what you like most. If there are certain mines you prefer, seek out the finest pieces you can afford. If it is an artist whose work you are drawn to, build a collection of their work. Regardless of what determination you use to collect or wear turquoise, when you take a deliberate rather than a scattershot approach, as you collect you will gain more expertise in that area of the gem, mine, or jewelry.

There are five recognized categories to rank turquoise, and the stone's value is in large part a function of its rank, as well as whether or not it is natural turquoise. It might surprise you to learn that less than one percent of all turquoise can justifiably be called a gem, though very high-grade stones are nearly gems.

Very high-grade turquoise indicates the matrix pattern, if any, is wholly balanced in the stone, or the color is not flawless throughout the piece, though this is minor in relation to the stone's overall quality.

High grade turquoise is hard, but may not be wholly balanced. The stone's luster is the highest quality. Often, this turquoise is used in the finest custom designed jewelry.

Jewelry grade turquoise has a good hardness and is usually not stabilized. Its luster is extremely good and the matrix, if any, is nicely balanced.

Good grade turquoise will take a good polish. It is pleasing in every respect, though of lesser quality than the higher grades. This turquoise almost always needs to be stabilized in order to be cut and polished and hold its color.

Chalk or bulk grade turquoise is the softest and most porous of all turquoise. It is extremely brittle and must be stabilized and often heat-treated or color enhanced to make it workable in any form.

Authenticity

The Indian Arts and Crafts Act (IACA) of 1990 (P.L. 101-644) implemented an important change benefitting the collector or consumer of Native American jewelry. The IACA prohibits misrepresentation in marketing of American Indian or Alaska Native arts and crafts products within the United States. It is illegal to offer or display for sale, or sell any art or craft product in a manner that falsely suggests it is Indian produced, an Indian product, or the product of a particular Indian or Indian tribe or Indian arts and crafts organization in the United States.

The law covers all Indian and Indian-style traditional and contemporary arts and crafts produced after 1935. The act broadly applies to the marketing of arts and crafts by any person in the United States. Some traditional items frequently copied by non-Indians include Indian-style jewelry, pottery, baskets, carved stone fetishes, woven rugs, kachina figures, and clothing.

The Indian Arts and Crafts Board, an agency established in 1934, has responsibility for overseeing the implementation of the Act.

The US Department of the Interior explicitly states that, "Under the Act, an Indian is defined as a member of any federally or State recognized Indian Tribe, or an individual certified as an Indian artisan by an Indian Tribe."

In Section 309.2, the Act defines an Indian tribe as:
1. Any Indian tribe, band, nation, Alaska Native village, or any organized group or community recognized as eligible for the special programs and services provided by the United States to Indians because of their status as Indians; or

2. Any Indian group that has been formally recognized as an Indian tribe by a state legislature or state commission or similar organization legislatively vested with a state tribal recognition authority.

All products must be marketed truthfully regarding the Indian heritage and tribal affiliation of the producers, so as not to mislead the consumer. It is illegal to market an art or craft item using the name of a tribe if a member, or certified Indian artisan, of that tribe did not actually create the art or craft item.

Section 309.4 of the Act also allows for individuals with tribal ancestry who are not eligible for enrollment to be designated as "an Indian artisan by a particular tribe."

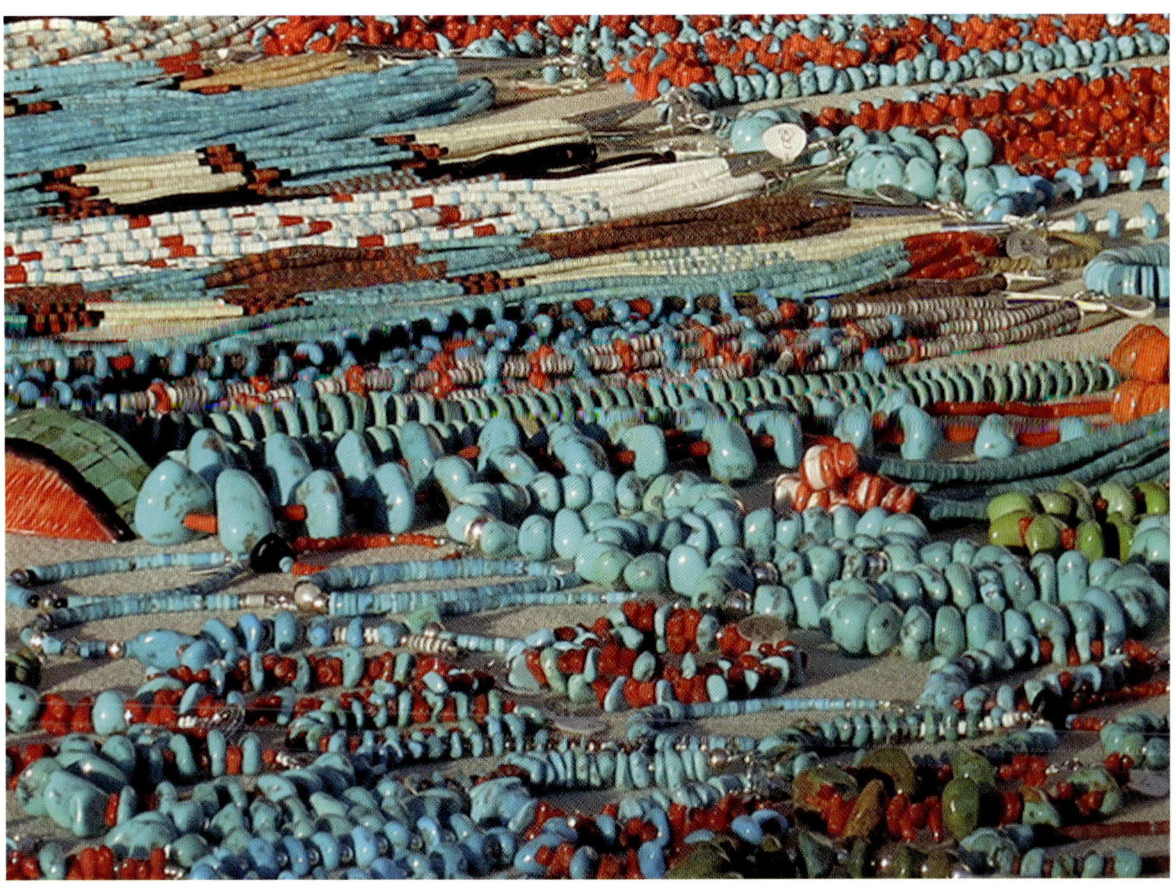

With an estimated 2.4 million visitors annually, the palace portal in Santa Fe has provided a reputable and reliable outlet for Southwest Native American turquoise jewelry and other arts and crafts for generations. *Courtesy of the author.*

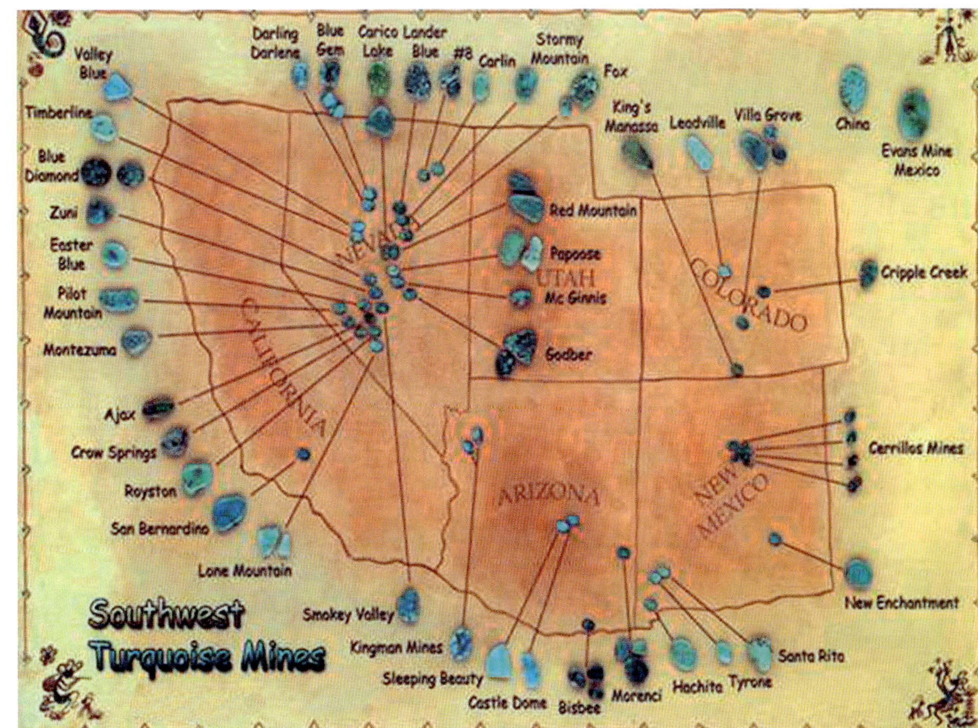

Turquoise mines map. Southwest, United States. *Courtesy of the author.*

CHAPTER 3

COUNTRIES OF ORIGIN

NATION upon nation can lay claim to being a host to turquoise in its native state. While some mines are extremely well-known and accessible, others are rare and often virtually impossible to gain entry to today. A number of countries produced vast quantities of various quality materials, while some nations mined small amounts that merit nothing more than a mere mention. This mention is important, however, in understanding the worldwide availability of turquoise and how trading routes came to be important in moving this prized gemstone from one region to another.

An interesting aside is that turquoise appears to have been considered a mineral of exceptional beauty, life, and good fortune, and disparate cultures around the world have attributed to it healing powers throughout the centuries.

The world map illustration, and the discussion of various countries where turquoise has been mined, serve as an important basis for understanding the availability of turquoise used in crafting.

As mentioned earlier, Egypt is recognized as the first nation to use turquoise for religious and ornamental purposes, going back as far as 5500 BCE. The mines in the Sinai were quite active during the rule of the ancient Egyptians, who firmly believed that turquoise had mystical and healing powers.

Persia holds a different distinction. The mines at Nishapur revealed the world's finest natural turquoise. The Persians were also the first people to use turquoise as a trade item, and bartering served them well. Ancient Persian turquoise has been found in various areas of Russia, including Siberia, so we know it traveled across the continent with earlier travelers or settlers, and graves in Turkistan have been opened to expose some of the finest cut Persian turquoise. Highly prized for its consistent, solid, robin's-egg-blue color, Persian turquoise continues to be the standard by which all other turquoise is judged.

Tibet has been a source of some of the most fantastic spider web turquoise for generations, though in limited quantities. Tibetans valued the gemstone above gold. Not only was turquoise used for personal ornamentation, it was also valued as a currency in many regions of the country.

China is today's leading producer of gem-quality turquoise. Dating to the thirteenth century, most of the early Chinese turquoise came from trade. In part because of the country's relatively closed society, we are aware of little from individual mines in China. While the turquoise can be of a high quality and is often mistaken for American turquoise, the Chinese have never given it the same vaunted status as jade.

South America was one of the earlier producers of turquoise as an ornamental and ceremonial gemstone. While few South American countries have major outcroppings, turquoise has been mined for centuries in small quantities in countries where the gem has a beautiful blue coloring. Peruvian turquoise is an excellent example. Having been overshadowed by its northern neighbor, the United States, little gem-grade material is associated with South America today.

United States turquoise has been the backbone of the gem cutter's craft for more than a century. Synonymous with the American Southwest, turquoise has been worked by Native Americans into some of the finest and most elaborate jewelry. It adorns those who favor the high quality and brilliant sky blue color that comes from a wonderfully set stone, whether in coin silver, sterling silver, or a silver and gold intarsia piece.

The jewelry made by Native Americans for spiritual use was often more elaborate and beautiful than that crafted strictly for ornamentation. This was more prevalent before today's burgeoning market for highly crafted silver and gold settings and finely detailed intarsia pieces. Much of the work being done by Native Americans of the Southwest these days can truly be considered fine art.

Arizona, Nevada, New Mexico, Colorado, and, to a lesser extent, California and Utah have been hosts to some of the most famous turquoise mines in history, where each vein yields a treasure trove of unique gems.

World map. *Courtesy of the author.*

Mining Claims

There are two types of mining claims (lode and placer), and two types of sites (mill site and tunnel site).

Lode claims include classic veins or lodes of deposits that have well-defined boundaries. They typically include other rock that can bear valuable minerals and may be broad zones of mineralized rock. Examples include quartz or other veins bearing gold or other metallic minerals and large-volume but low-grade disseminated metallic deposits. Lode claims are usually described as parallelograms with the longer side lines parallel to the vein or lode. Federal statute limits their size to a maximum of 1,500 feet in length along the vein or lodge. Their width is a maximum of 600 feet, 300 feet on either side of the center line of the vein or lode. The end lines of the lode claim must be parallel to qualify for underground extralateral rights. Extralateral rights involve the rights to minerals that extend at a depth beyond the vertical boundaries of the claim.

Placer claims are mineral deposits not subject to lode claims. Originally, these included only deposits of unconsolidated materials, such as sand and gravel that contained free gold or other minerals. By congressional acts and judicial interpretations, many non-metallic bedded or layered deposits, such as gypsum and high calcium limestone, are also considered placer deposits. The maximum size of a placer claim is twenty acres per locator.

Mill site must be located on non-mineral land. Its purpose is to either support a lode or placer mining claim operation or support itself independent of any specific claim. A mill site must include the erection of a mill or reduction works and/or may include other uses reasonably practical to the support of a mining operation. The maximum size of a mill site is five acres.

Tunnel site is where a tunnel is run to develop a vein or lode. It may also be used for the discovery of unknown veins or lodes. To stake a tunnel site, two stakes are placed up to 3,000 feet apart on the line of the proposed tunnel. Record keeping is the same as a lode claim.

An individual may locate lode claims to cover any veins or lodes intersected by the tunnel, including any not seen. The maximum distance for these lode claims is 1,500 feet on either side of the center line of the tunnel, thus giving the mining claimant the right to prospect an area 3,000 feet wide and 3,000 feet long.

CHAPTER 4

MINING LOCALES

TURQUOISE has been extracted from a wide variety of mines, some well-known and others obscure or forgotten. Both types are included here, as it is important for collectors to understand the historical context and value these localities. Their inclusion is intended to paint a broad brushstroke of the multitude of mines from which turquoise has been worked, and the following images highlight some of the most beautiful turquoise to be found anywhere in the world.

Ajax (Nevada) mine is one of the smaller mines in south-central Nevada. In close proximity to the better-known Royston mining area, the Ajax mine is one of a relatively few newer American turquoise mines. The material mined here ranges from light blue with blue veins to an overall dark green stone with light blue coloring.

Apache Canyon (California) turquoise has been mined since the time of the Anasazis. In modern days the mine was rediscovered in the 1960s and is still in production on a limited basis. The turquoise from Apache Canyon is typically a beautiful, light robin's-egg-blue with golden yellow matrix in the open web pattern.

Australian (Australia) produces a limited amount of high-quality turquoise, in Victoria and South Wales, in the Southeast part of the continent. Approximately twenty Australian mines have reported turquoise deposits in varying quantities, though the quality varies greatly from mine to mine.

Ajax turquoise. Necklace. Multiple stone beads. Pueblo. 18". *JR's Southwestern Trading Post.*

Australian turquoise. Harts Ranges, Northern territory, Australia. This area became mined out in 1952. All material in this image is natural with no backing. *Photo courtesy of Boris Branwhite.*

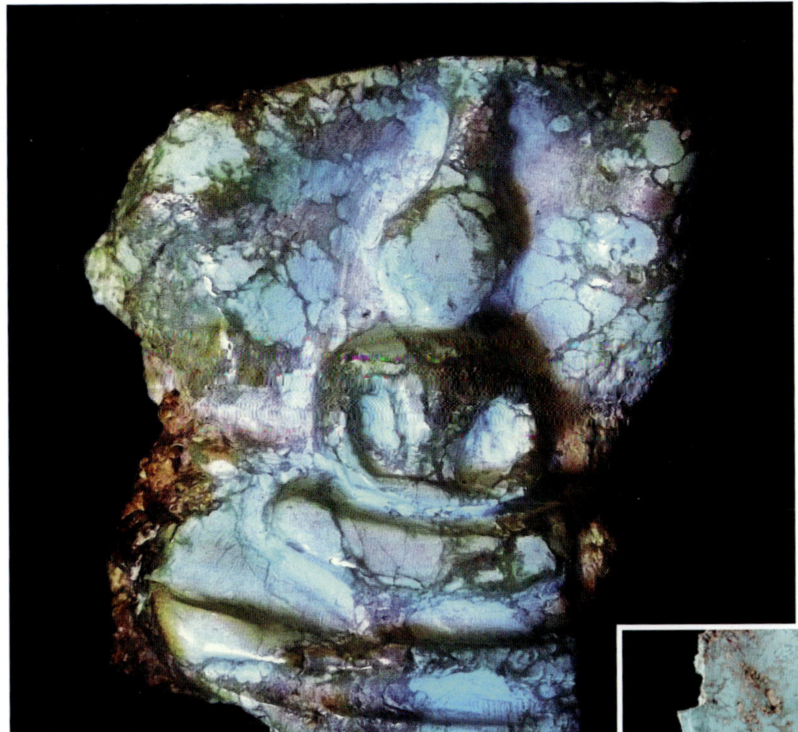

Australian turquoise. Carving. In the 1940s, the Harts Range Phosphate deposits were mined. Amongst the phosphate, turquoise was found, put into drums, and stored on site. The mine closed in 1952, and was not rehabilitated. In the early 1970s, during the $2 shelf company rage, one of Alan Bond's shelf companies purchased the lease and cleaned it up. The drums of turquoise were sold and shipped to Sydney. The new owner put them into a vault where they remained until his death in 1986, then sold by auction to a local fossicker/miner/mineral collector. The material remaining is the last of this mine-run collection. The patterns produced within this material mimic many of the famous Southwest American turquoise mines, such as #8, Carico Lake, Candelaria, Royston, Kingman, Pilot Mountain, and others. *Photo courtesy of Boris Branwhite.*

Australian turquoise. Slabs. *Photo courtesy of Boris Branwhite.*

Australian turquoise. Seashell internal replacement. Shell sitting on a bed of turquoise. *Photo courtesy of Boris Branwhite.*

Bisbee (Arizona) is named for its location in Bisbee. This mine produces one of the most famous types of turquoise in the world, and its extraction is part of the larger Bisbee copper mine operation. The turquoise is recognized for its relative hardness and wonderful webbing, along with brilliant blue coloring. The matrix that forms the incredible natural spider web design has been called "Smokey Bisbee." Bisbee turquoise is found at all levels of the mine. Though the highest grade is from the one-hundred-foot level, good-quality stones are found as high as 2,000 feet. The Phelps Dodge Mining Company, the mine's owner, announced that Bisbee is nearly depleted, and the mine has been buried under tons of dirt. Bisbee will always be prized for its rarity, spider webbing, and natural beauty.

Bisbee turquoise. Belt buckle. Multiple stones. Silver setting. Navajo, signed Justin Moris. 4" × 3". *JR's Southwestern Trading Post.*

Bisbee turquoise. Assorted cut and polished stone. 24 grams. *Courtesy of John Renner.*

Bisbee turquoise and coral. Assorted belt buckles. Multiple stones and chip inlay. Silver setting. Navajo. 2¼" × 2½", 2" × 1¾", 2" × 1¾". *JR's Southwestern Trading Post.*

Bisbee II turquoise. Assorted cut and polished stone. 16 grams. *Courtesy of John Renner.*

Bisbee turquoise. Cuff bracelet. Multiple stones. Silver setting. Navajo. 2¼" × 3½. *JR's Southwestern Trading Post*

Bisbee turquoise. Pendant. Silver setting. Navajo. 2¾" × 2¼". *JR's Southwestern Trading Post.*

Bisbee turquoise. Squash blossom necklace. Multiple stones. Silver setting. Navajo. 24". *JR's Southwestern Trading Post.*

Bisbee turquoise. Vintage necklace and shadowbox cuff bracelet. Silver setting. Blackwater Trading Post collection. Loose cabochons, Western Trading Post. *Photo courtesy of Western Trading Post.*

Bisbee turquoise. Cuff bracelet, ring, and earrings. Silver setting. Navajo, signed Tommy Jackson. Western Trading Post. *Photo courtesy of Western Trading Post.*

Bisbee turquoise. Cuff bracelet. High grade twist-wire. Silver setting. C. 1960s. *Courtesy of John Miller.*

Bisbee turquoise. Bisbee (Persian beads) on slab of Bisbee turquoise. Silver setting. *Courtesy of John Miller.*

Bisbee turquoise. Cuff bracelet. Tufa cast. Scene depicts the Navajo's "Long Walk" (1864–66). Silver setting. Navajo, signed Philander Begay. *Courtesy of John Miller.*

Bisbee turquoise. Foliage Pendant. Silver setting. C. 1970s. *Courtesy of John Miller.*

Bisbee turquoise. Cuff bracelet. High-grade black matrix. Silver setting. Navajo, signed Jimmy Yazzie. C. 1980s. *Courtesy of John Miller.*

Bisbee turquoise. Bracelet. "Birtha Bisbee." One-of-a-kind natural Bisbee pictorial with a geologically accurate depiction of Bisbee mining region. Silver setting. Navajo, signed Eddy Chaco. C. 1980s. *Courtesy of John Miller.*

Bisbee turquoise. Cuff bracelet. Silver setting. *Courtesy of John Miller.*

Deep stamp cuff bracelets (top to bottom): Bisbee turquoise. Silver setting. Navajo, signed Delbert Gordon; Blue Gem turquoise and coral. Navajo, signed Ernest Roy Begay (ERB); Bisbee turquoise. Navajo, signed Jimmy a Yazzie; Bisbee turquoise. Navajo, signed Delbert Gordon. *Courtesy of John Miller.*

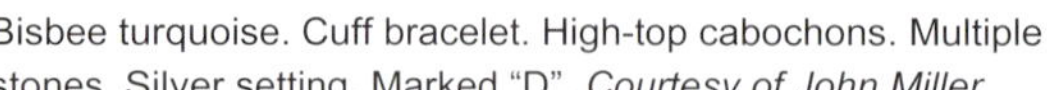

Bisbee turquoise. Cuff bracelet. High-top cabochons. Multiple stones. Silver setting. Marked "D". *Courtesy of John Miller.*

Bisbee turquoise. Cuff bracelet, necklace, ring, earring set. Lavender Pit. Multiple stones. Silver setting. C. 1980s. *Courtesy of John Miller.*

Bisbee turquoise. Cuff bracelet. Silver setting. Navajo, signed Charlie John. *Courtesy of John Miller.*

Bisbee turquoise. Rings. "His and Hers," butterfly-cut. Silver setting. Navajo, signed Tommy Jackson. *Courtesy of John Miller.*

Bisbee turquoise. Cuff bracelet. (inside) Deep stamp-work. Silver. Navajo, signed Delbert Gordon. *Courtesy of John Miller.*

Bisbee turquoise. Cuff bracelet. Deep stamp-work. Multiple stones. Silver setting. Five consecutive slices cut from one single turquoise. 50 carats. Navajo, signed Delbert Gordon. *Courtesy of John Miller.*

Bisbee turquoise. Spiderweb turquoise with quartz crystal spikes protruding into the face of the stone. Vein rock sandwiched between two thick layers of quartz with light fractures. 65 carats. *Courtesy the collection of Michael Griffo.*

Bisbee turquoise. Cabachon. "Smokey" turquoise. 65.5 carats. *Courtesy the collection of Michael Griffo.*

Bisbee turquoise. Breccia matrix cabochon. *Courtesy of John Miller.*

Bisbee turquoise. Rings. Silver setting. Top and bottom two left, Michael Griffo; top right, Navajo, signed Kee Yazzie; bottom right, unknown silversmith. *Courtesy the collection of Michael Griffo.*

Bisbee turquoise. Cuff bracelet. Wire-twist. Silver setting. Navajo, signed Darrell Cadman. *Courtesy of John Miller.*

Bisbee turquoise. Cuff bracelet. Deep stamp-work. Multiple stones. Silver setting. Navajo, signed Delbert Gordon. 100 carats. *Courtesy of John Miller.*

Bisbee turquoise. Cuff bracelet. (inside) Deep stamp-work. Silver. Navajo, signed Delbert Gordon. *Courtesy of John Miller.*

Bisbee turquoise. Cut slab. 1'+. *Courtesy of John Miller.*

Bisbee turquoise. Squash blossom necklace. Multiple stones. Silver setting. C. 1970s. *Courtesy of John Miller.*

Bisbee turquoise. Cuff bracelet. Multiple stones. Silver setting. Seventy-five carats. Total weight, 5.8 ounces. Navajo, signed Erick Begay. *Courtesy the collection of Michael Griffo.*

Bisbee turquoise. Persian beads, bracelets, rings. Multiple stones. Silver setting. *Courtesy of John Miller.*

Blue Bird (Arizona) mine is a little-known mine in Cochise County. Turquoise from this mine is typically dark blue with quartz and black matrix. The gemstone is hard, solid material and can also be found in a medium blue color with a yellow-brown matrix. Typically smaller pieces are worked into cabochons for setting. The mine is not known for producing large workable materials.

Blue Diamond (Nevada) mine opened in the late 1950s and was worked until 1980. The material from this mine was typically plate form, and often gave the appearance of Stormy Mountain turquoise due to its black matrix. The vibrant blue of the mineral is contrasted by the smoky matrix, though colors can range from blue to green. The mine, which was buried under tons of rock, is no longer easily accessible; however, the occasional miner will dig the area on a small scale to uncover plate formations.

Blue Diamond turquoise. Cuff bracelet. Silver setting. C. 1970s. Unmarked. *Courtesy of John Miller.*

Blue Diamond turquoise. Squash blossom necklace and cuff bracelet. Multiple stones. Silver setting. C. 1970s. *Courtesy of John Miller.*

Blue Gem (Nevada) mine is regarded as one of Nevada's most prolific mines for the immense variety of turquoise recovered. Found near Battle Mountain, the turquoise of Blue Gem runs from distinctively deep blue to deep green in matrix. Blue Gem should not be confused with other Nevada mines of the same name. The Battle Mountain Blue Gem mine is now closed, though the material is still highly sought after by jewelers for its rich color.

Blue Gem turquoise. Belt buckle. Multiple stones. Silver setting. Navajo, signed ew. 4¾" × 3". *JR's Southwestern Trading Post.*

Blue Gem turquoise. Squash blossom necklace. Multiple stones. Silver setting. Navajo. 26". *JR's Southwestern Trading Post.*

Blue Gem turquoise. Necklace. Multiple stones and beads. 22". *JR's Southwestern Trading Post.*

Blue Gem turquoise. Concho belt. Multiple stones. Silver setting. 2½" × 2" (buckle), 2¼" × 1¾" (Concho). *JR's Southwestern Trading Post.*

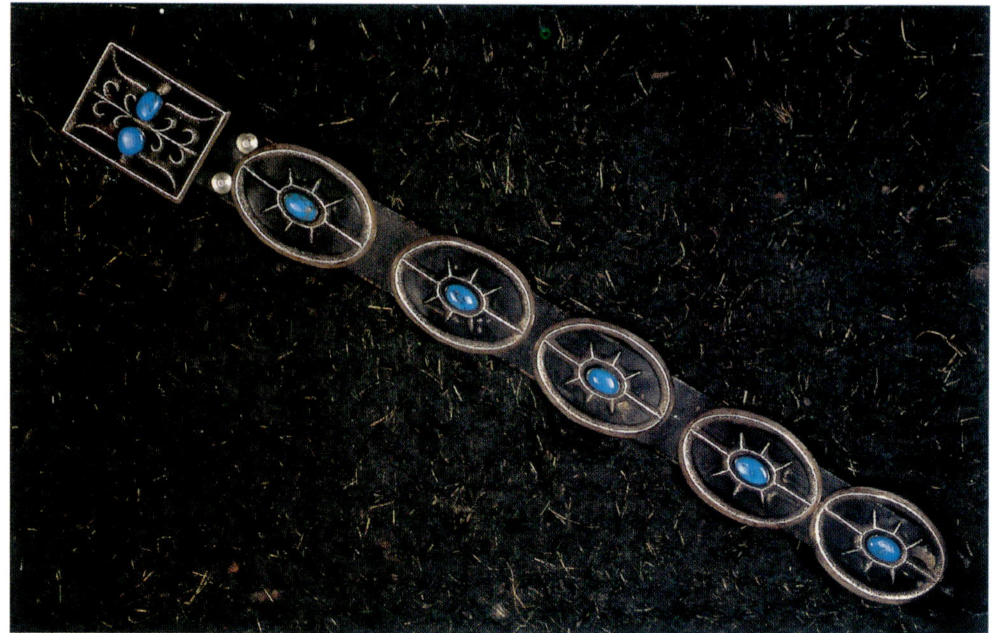

Blue Gem turquoise. Cuff bracelet. Silver setting. Unmarked. C. 1940s–1950s. *Courtesy of John Miller.*

Blue Gem turquoise. Bracelet. Turquoise and coral. Deep-stamped. Multiple stones. Silver setting. Navajo, signed Ernest Roy Begay (ERB). *Courtesy of John Miller.*

Blue Gem turquoise. Watch cuff bracelet. Multiple stones, inlay. Silver setting. Navajo. 1¼" × 3". *JR's Southwestern Trading Post.*

Blue Gem turquoise. Cuff bracelet. Multiple stones, cluster. Silver setting. Navajo. 3" × 3". *JR's Southwestern Trading Post.*

Blue Gem turquoise. Cuff bracelet. Silver setting. Navajo, signed ES. 1" × 3". *JR's Southwestern Trading Post.*

Blue Gem turquoise. Cuff bracelet. Silver setting. Navajo. 2" × 3". *JR's Southwestern Trading Post.*

Blue Gem turquoise. Cuff bracelet. Multiple stones. Silver setting. Navajo. 1½" × 3". *JR's Southwestern Trading Post.*

Blue Gem turquoise. Cuff bracelet. Multiple stones. Silver setting. 2" × 3". *JR's Southwestern Trading Post.*

Blue Gem turquoise. Pendant necklace. Harvey-style. Multiple stones. Silver setting. Unmarked. C. 1970s. *Courtesy of John Miller.*

Blue Gem turquoise. Cuff bracelet. Silver setting. C. 1940–1950. *Courtesy of John Miller.*

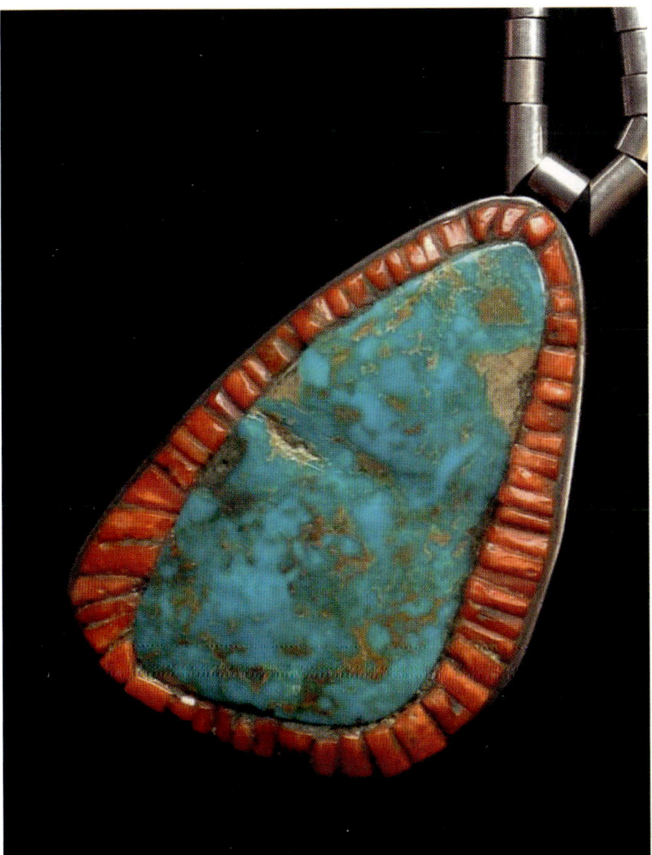

Blue Gem turquoise. Pendant necklace. Turquoise and coral. Silver setting. Unmarked. *Courtesy of John Miller.*

Battle Mountain turquoise (Blue Gem turquoise region). Concho belt. Multiple stones. Silver setting. *Courtesy of John Miller.*

Blue Jay turquoise. Belt buckle. Multiple stones. Silver setting. Navajo. 3½" × 2½". *JR's Southwestern Trading Post.*

Blue Jay (Nevada) mine is in the state's Battle Mountain region. It is one of the smaller Nevada mines that produces large, high-grade gem nuggets. Its blue color is similar to turquoise from the Fox mine. This turquoise may also have large spider webbing or brown matrix in the cut material. It is an excellent turquoise to cut and polish because of its relative hardness, and it looks stunning when set in silver as a pendant or in a bracelet.

Candelaria (Nevada) mine is another of the now depleted and closed Nevada mines that produced gem-grade, rich blue stone, occasionally with brown or black matrix. Collectors know it for its beautiful quality and rarity.

Candelaria turquoise. Cuff bracelet. Silver setting. Navajo. ¾" × 3". *JR's Southwestern Trading Post.*

Candelaria turquoise. Cuff bracelet. Multiple stones. Gold setting. Signed John Renner. 1" × 3". *JR's Southwestern Trading Post.*

Candelaria turquoise. Cluster ring. Multiple stones. Silver setting. C. 1950–1960. *Courtesy of John Miller.*

Carico Lake (Nevada) mine is a dry lake bed in Lander County. Recognized for the beauty of the area, Carico Lake is at a higher elevation than many turquoise mines. While Carico Lake is renowned as a gold-producing mine, the mining company has leased the turquoise-producing area to miners and prospectors, and fine, gem-grade material is recovered from time to time. Because of the limited area and time to work a claim, the turquoise is a beautiful addition to any collection. You'll find some outstanding Carico Lake turquoise primarily in Irish green colors. This is due in large measure to the zinc content in the material. Even rarer is the dark blue-green Carico Lake turquoise with exceptional spider web matrix.

Carico Lake turquoise. Assorted cut and polished stones. 18 grams. *Courtesy of John Renner.*

Carico Lake, Kingman, Persian, and Lone Mountain turquoise. Rings. Gold and silver settings. *JR's Southwestern Trading Post.*

Carico Lake turquoise. Cuff bracelet. Multiple stones. Silver setting. Navajo. 1" × 3". *JR's Southwestern Trading Post.*

Carico Lake turquoise. Bolo tie. Multiple stones. Silver setting with silver tips. Navajo, signed AZ. 3½" × 2½". *JR's Southwestern Trading Post.*

Carico Lake turquoise. Cuff bracelet. Multiple stones. Silver setting. Navajo, signed Verdy Jake. *Courtesy of John Miller.*

Carico Lake turquoise. Bolo tie. Multiple stones. Silver setting. Signed. *Courtesy of the author.*

Carico Lake turquoise. Necklace. Turquoise and coral beads. Navajo. 24". *JR's Southwestern Trading Post.*

Carlin (Nevada) mine is in Nevada's northeast quadrant. It is very small and no activity has taken place in many years. Carlin produced a very hard turquoise of exceptional clarity. Much of the material is gel type, more like silica with a distinctive look. It has been confused at times with Stormy Mountain turquoise; however, these are two separate mines worked by different miners.

Castle Dome (Arizona) mine is an open pit copper mine from which beautiful, mostly solid blue turquoise was mined. Turquoise, copper, and other minerals often share mining areas because of their similar chemical make-up. Castle Dome turquoise is a good example of copper mining revealing stunning, gem-grade stones.

Carlin turquoise. Necklace beads. *Photo courtesy of Western Trading Post.*

Castle Dome turquoise. Belt buckle. Multiple stones, cluster. Silver setting. Zuni, signed W. V. Quam. 4" × 3". *JR's Southwestern Trading Post.*

Cave Creek (Arizona) mine is one of the relatively newer mining operations and is located outside of Cave Creek, Arizona. Operated by a father and son, the material is a medium to dark blue color; some have compared it to older Kingman or Morenci turquoise. Often the material has pyrite inclusions, and while there has not been much material mined to date, it is sought after by collectors and artists.

Cerrillos (New Mexico) mine turquoise is not only rich in beauty, but also in history. The oldest known mine of any kind in North America, Cerrillos is situated only ten miles from Santa Fe, New Mexico. With a large deposit of turquoise, some is partially exposed and therefore easier to work. The early miners used stone axes, chisels, and picks to pry the precious material from its host rock. It is believed that Pueblo miners, who worked the mine the most, recovered more than 100,000 tons of rock, creating a mine pit some 200 feet deep. In addition, vertical shafts were also dug to reach turquoise veins. Cerrillos turquoise was used as a trade tool of these earlier people, and the Pueblos continued mining the mineral through the early 1870s. The silver boom was on, and interest in the area was heightened when Tiffany's of New York bought the mined area and recovered more than $2 million worth of turquoise in the 1890s. One of North America's most workable mined turquoise, Cerrillos is synonymous with high-quality and gem-grade stones. Because of its relative hardness, Cerrillos turquoise is often easier to work with and takes a more lustrous polish than the stones mined at other locations. Having been formed at the base of a volcano, Cerrillos turquoise was created under unusual circumstances. These unique conditions have given Cerrillos a plethora of colors created from other minerals in the numerous host rocks. More than seventy-five color shades have been identified, ranging from rather bland yet rare light brown or tan to extremely vibrant blue-green gem tones.

Cerrillos turquoise. Assorted cut and polished stones. 56 grams. *Courtesy of John Renner.*

Cerrillos turquoise. Pins. Multiple stones. Silver setting. Navajo. 2¾" × 1¼", 2¼" × 2". *JR's Southwestern Trading Post.*

Cerrillos turquoise. Pin. Silver setting. Navajo. 1¾" × ½". *JR's Southwestern Trading Post.*

Cerrillos turquoise. Cuff bracelet. Multiple stones. Silver setting. 2" × 3". *JR's Southwestern Trading Post.*

Cerrillos turquoise. Cuff bracelet. Silver setting. 2" × 3". *JR's Southwestern Trading Post.*

Cerrillos turquoise. Cuff bracelet. Turquoise and coral. Multiple stones. Silver setting. Navajo. 1" × 3". *JR's Southwestern Trading Post.*

Cerrillos turquoise. Cuff bracelet. Silver setting. Navajo. 1" × 3". *JR's Southwestern Trading Post.*

Cerrillos turquoise. Cuff bracelet. Multiple stones. Silver setting. Signed John Renner. 2" × 3½". *Courtesy of John Renner.*

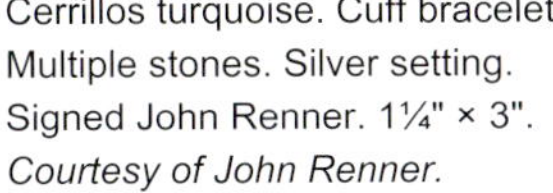

Cerrillos turquoise. Cuff bracelet. Multiple stones. Silver setting. Signed John Renner. 1¼" × 3". *Courtesy of John Renner.*

Chinese (China) turquoise accounts for eighty percent of all turquoise on the US market. Nearly all Chinese turquoise has been stabilized to harden the stone before being set into jewelry. The majority of Chinese turquoise being mined comes from the Hubei Province that produces turquoise similar to that of depleted Southwest US mines, particularly those in Nevada, to the confusion of collectors and jewelers. Other deposits have been worked from the Ma'ashan mine northwest of the large city of Shanghai. In the Academy of Social Sciences in Bejing is a bronze plaque with turquoise overlay dated to the Erlitou site, giving rise to the belief that turquoise mining dates to 1700 B.C.E. in China. Most early turquoise in the region is credited to trade between the Chinese, Persians, Turks, and others. Specific information on today's mining operations today is not readily available, due in part to governmental controls.

Chinese turquoise. Assorted cut and polished stones. 46 grams total. *Courtesy of John Renner.*

Chinese turquoise. Belt buckle. Silver setting. Navajo, signed H. B. 3½" × 3". *JR's Southwestern Trading Post.*

Chinese turquoise. Concho belt. Multiple stones. Silver setting. 3" × 2½", 2½" × 2". *JR's Southwestern Trading Post.*

Chinese turquoise. Belt buckles. Silver settings. Navajo. 3" × 2½". *JR's Southwestern Trading Post.*

Chinese turquoise. Necklace. Assorted beads. Navajo. 24". *JR's Southwestern Trading Post.*

Chinese turquoise. Necklace. Multiple stones, double strand beads. Navajo. 24". *JR's Southwestern Trading Post.*

Chinese turquoise. Necklace. Coral and shell, turquoise beads. Multiple stones, double strand. Navajo. 20". *JR's Southwestern Trading Post.*

Chinese turquoise. Necklace. Turquoise and coral. Multiple stones, eight strand. Navajo. 22". *JR's Southwestern Trading Post.*

Chinese turquoise. Belt buckle. Silver setting. Navajo, signed H. B. 3½" × 3". *JR's Southwestern Trading Post.*

Chinese turquoise. Pin. Silver setting. Navajo. 2½" × 2". *JR's Southwestern Trading Post.*

Chinese turquoise. Necklace. Multiple stone, three strands. Pueblo. 24". *JR's Southwestern Trading Post.*

Chinese turquoise. Belt buckle. Silver setting. Navajo, signed VY. 4½" × 3". *JR's Southwestern Trading Post.*

Chinese turquoise. Pendant. Turquoise, coral, lapis. Multiple stones, inlay. Silver setting. Navajo, signed L. 2" × 1". *JR's Southwestern Trading Post.*

Chinese turquoise. Pendant. Silver setting. Navajo, signed Santa Fe DEU. 2" × 1¾". *JR's Southwestern Trading Post.*

Chinese turquoise. Pendant. Multiple stones. Silver setting. Navajo. 2" × 1". *JR's Southwestern Trading Post.*

Chinese turquoise. Pendant. Multiple stones, cluster. Silver setting. 2" × 1". *JR's Southwestern Trading Post.*

Chinese turquoise. Pin. Multiple stones. Silver and brass setting. Navajo. 2½". *JR's Southwestern Trading Post.*

Chinese turquoise. Necklace. Multiple stone nuggets. Navajo. 12". *JR's Southwestern Trading Post.*

Chinese turquoise. Earrings. Multiple stones, cluster. Silver setting. Navajo. 1½". *JR's Southwestern Trading Post.*

Chinese turquoise and coral. Necklace. Turquoise and coral, five strands. Navajo. 20". *JR's Southwestern Trading Post.*

Chinese turquoise. Pendant. Silver setting. Navajo, signed CM. 2½" × 3¼". *JR's Southwestern Trading Post.*

Chinese turquoise. Pendant. Bead necklace. Silver setting. Navajo, signed DB. 20". *JR's Southwestern Trading Post.*

Chinese turquoise. Pendant. Silver setting. Navajo, signed D. 3¼" × 1¾". *JR's Southwestern Trading Post.*

Chinese turquoise. Pendant. Bead necklace. Silver setting. Navajo, signed DB. 20". *JR's Southwestern Trading Post.*

Chinese Turquoise. Pendants. Silver settings. Navajo, signed DB. 1" × ¾", 1¼" × 7/8", 1" × 1". *JR's Southwestern Trading Post.*

Chinese turquoise. Pendant. Silver setting. Navajo, signed E. 2½" × 1". *JR's Southwestern Trading Post.*

Chinese turquoise. Pendant. Silver setting. Navajo. 1¾" × 1½". *JR's Southwestern Trading Post.*

Chinese turquoise. Pin. Silver setting. Navajo. 3 × 2¾". *JR's Southwestern Trading Post.*

Chinese turquoise. Pendant. Silver setting. Navajo, signed E. 2" × 1". *JR's Southwestern Trading Post.*

Chinese turquoise. Pin. Silver setting. 1½" × 1½". *JR's Southwestern Trading Post.*

Chinese turquoise. Belt buckle. Silver setting. Brad Davis. 4" × 4". *JR's Southwestern Trading Post.*

Chinese turquoise. Pendant. Silver setting, liquid silver strands. Navajo, signed Elouise. 12". *JR's Southwestern Trading Post.*

Chinese turquoise. Pendant. Silver setting. Navajo, signed H. 2½" × 2". *JR's Southwestern Trading Post.*

Chinese turquoise. Pendant. Silver setting. Navajo, signed DB. 2½" × 1¼". *JR's Southwestern Trading Post.*

Chinese turquoise. Pendant. Silver setting. 2¼" × 2". *JR's Southwestern Trading Post.*

Chinese turquoise. Cuff bracelet. Silver setting. Navajo. 2½" × 2". *JR's Southwestern Trading Post.*

Chinese turquoise. Pendant. Silver setting. Navajo, signed AS. 2½" × 1". *JR's Southwestern Trading Post.*

Chinese turquoise. Pendant. Silver setting. Navajo, signed E. Etsity. 4¾" × 3¼". *JR's Southwestern Trading Post.*

Chinese turquoise. Pendant. Silver setting. 1¾" × ¾". *JR's Southwestern Trading Post.*

Chinese turquoise. Cuff bracelet. Silver setting. Navajo. 3" × 3". *JR's Southwestern Trading Post.*

Chinese turquoise. Cuff bracelet. Silver setting. Navajo. 4" × 3½". *JR's Southwestern Trading Post.*

Chinese turquoise. Pendant. Silver setting. Navajo, signed DB. 2" × 1". *JR's Southwestern Trading Post.*

Chinese turquoise. Pendant. Silver setting. Signed John Renner. 4¾" × 3". *Courtesy of the artist.*

Chinese turquoise. Pendant. Silver setting. Navajo, signed DB. 3½" × 1¼". *JR's Southwestern Trading Post.*

Chinese turquoise. Necklace. Multiple stones. Silver setting. Navajo. 28". *JR's Southwestern Trading Post.*

Chinese turquoise. Necklace. Turquoise and coral beads. 20". *JR's Southwestern Trading Post.*

Chinese turquoise. Cuff bracelet. Turquoise, coral and shell. Multiple stones, inlay. Silver setting. 2" × 3". *JR's Southwestern Trading Post.*

Side view.

Chinese turquoise. Cuff bracelet. Turquoise and coral Multiple stones, cluster. Silver setting. Navajo, signed Kirt Smith. 3½" × 3". *JR's Southwestern Trading Post.*

Chinese turquoise. Cuff bracelet. Multiple stones. Silver setting. Navajo, signed DM. 1¼" × 3". *JR's Southwestern Trading Post.*

Chinese turquoise. Cuff bracelet. Multiple stones. Silver setting. Signed Nakai. 1" × 3". *JR's Southwestern Trading Post.*

Chinese turquoise. Cuff bracelet. Silver setting. Navajo. 3½" × 3". *JR's Southwestern Trading Post.*

Chinese turquoise. Cuff bracelet. Multiple stones, inlay. Silver setting. Navajo, signed Lon Parker ¾" × 3". *JR's Southwestern Trading Post.*

Chinese turquoise. Cuff bracelet. Multiple stones. Silver setting. Signed Tom Dewit. 1" × 3". *JR's Southwestern Trading Post.*

Chinese turquoise. Cuff bracelet. Multiple stones, cluster. Silver setting. 2½" × 3". *JR's Southwestern Trading Post.*

Chinese turquoise. Cuff bracelet. Coral, lapis, and black onyx. Multiple stones. Silver setting. Navajo. 2½" × 3". *JR's Southwestern Trading Post.*

Chinese turquoise. Cuff bracelet. Silver setting. Navajo, signed Padilla. 2¾" × 3". *JR's Southwestern Trading Post.*

Chinese turquoise. Pins. Assorted gems. Gold settings. Signed E.S. 3½" × 3½". *JR's Southwestern Trading Post.*

Chinese turquoise. Pins. Assorted gems. Silver settings. Signed E.S. 3" x 3¼". *JR's Southwestern Trading Post.*

Chinese turquoise. Cuff bracelet. Multiple stones. Silver setting. Navajo, signed ES. ¾" × 3". *JR's Southwestern Trading Post.*

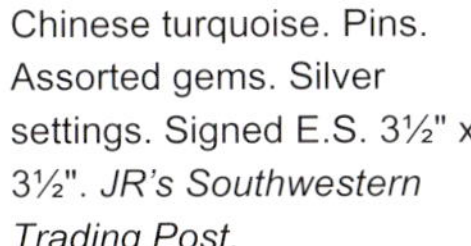

Chinese turquoise. Pins. Assorted gems. Silver settings. Signed E.S. 3½" x 3½". *JR's Southwestern Trading Post.*

Chinese turquoise. Pin. Assorted gems. Silver setting. Signed E.S. 3½" × 3½". *JR's Southwestern Trading Post.*

Chinese turquoise. Butterfly cuff bracelet. "Papillon." Silver setting. Navajo, signed Delbert Gordon. *Courtesy of John Miller.*

Chinese turquoise. Bracelet. Multiple stones, channel inlay. Silver setting. Navajo, singed A. Francisco. *Courtesy of John Miller.*

Chinese turquoise. Cuff bracelet. "Star" drop-button. Silver setting. Navajo, signed Calvin Martinez. *Courtesy of John Miller.*

Chinese turquoise. Cuff bracelet. Multiple stones. Silver setting. *Courtesy of John Miller.*

Chinese and Sleeping Beauty turquoise. Pendant and cuff bracelet. Multiple stones, multi-chromatic modern inlay. Silver setting. Navajo, signed Wayne Musket. *Courtesy of John Miller.*

Chinese turquoise. Cuff bracelet. Multiple stones. Silver setting. *Courtesy of John Miller.*

Chinese turquoise. Pendant. Silver setting. *Courtesy of John Miller.*

Chinese turquoise. Ring. Silver setting. Navajo, signed Kevin Barnhill. *Courtesy of John Miller.*

Chinese turquoise. Cuff bracelet. Seamless inlay. Silver setting. Zuni, signed J. Coonsis. *Courtesy of John Miller.*

Cripple Creek (Colorado) is known primarily as a gold-producing site. In a number of small mines, turquoise has been found as a by-product of gold mining. Cripple Creek turquoise is usually green, though there has been some blue with matrix material found as well. Cut and polished stones from this area are a rare addition to any collection.

Damele (Nevada) mine is close to the Carico Lake mine. It produces a distinctive hard, yellow-green stone largely due to the high zinc content. It is most often found with spider webbing of dark brown or black, and quantities are limited because of the mine's small size. Its rare yellowish turquoise is highly collectible.

Darling Darlene (Nevada) mine turquoise was discovered in the early 1970s. Darling Darlene turquoise ranges in color from greens to light to dark blue. This small mine, worked throughout the 1970s, was named after the claim owner's daughter passed away at a young age. The material has been worked into beautiful cabochon stones but has become increasingly difficult to find.

Cripple Creek turquoise. Watch cuff bracelet. Multiple stones. Silver setting. Navajo. 1½" x 3". *JR's Southwestern Trading Post.*

Damele turquoise. Cuff bracelet. Multiple stones. Silver setting. John Renner. ¾" x 3". *JR's Southwestern Trading Post.*

Damele turquoise. Concho belt. Multiple stones. Silver setting. Navajo. 3¾" x 3¼". *Courtesy of the artist.*

Detail.

Dry Creek (Nevada) has been known by a few names, including Godber-Brunham, as it is found on the claim. It is one of the Nevada mines that yielded turquoise of creamy white to pale blue. Well-known to collectors and specialists for its inviting white coloring, the turquoise is quite hard and its color is the result of an unusually high amount of aluminum in relation to copper content in the stone. The matrix is a pleasing light yellow, gray ,or black color. This turquoise is highly collectible, though you will find it most often set in jewelry with more traditional blue stones.

Dry Creek turquoise. Cuff bracelet. Multiple stones. Silver setting. Signed John Renner. 1½" x 3". *JR's Southwestern Trading Post.*

Easter Blue (Nevada) mine turquoise is some of the oldest to come out of Nevada. The mine is in the Royston area and has produced gem-grade blue-green stones with a matrix. A classic turquoise example from this region, Easter Blue sets well alone or with other stones.

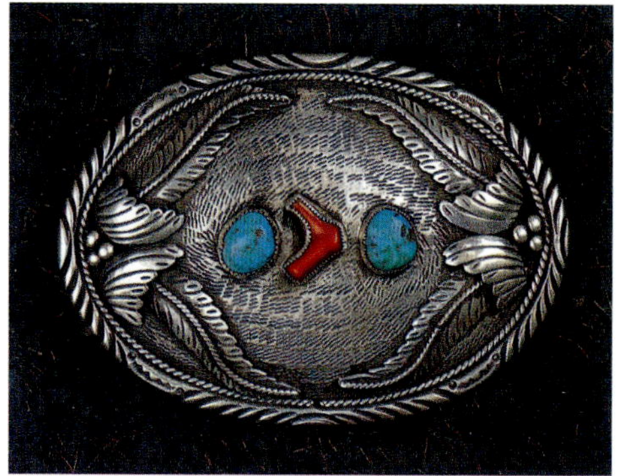

Easter Blue turquoise. Belt buckle. Turquoise and coral. Multiple stones. Navajo. 5" x 3½". *JR's Southwestern Trading Post.*

Easter Blue turquoise. Assorted cut and polished stones. 54 grams. *Courtesy of John Renner.*

Easter Blue turquoise. Cuff bracelet. Multiple stones. Silver setting. Navajo. 2½" x 3". *JR's Southwestern Trading Post.*

Side view.

Emerald Valley (Nevada) mine turquoise is much as its name implies, a beautiful lush green. The mine is still worked and the stone is usually heavily mixed with matrix of light tan to brown. A unique turquoise like Emerald Valley is outstanding for use in cut and polished jewelry.

Enchantment (New Mexico) mine was discovered by a gold miner in 1996 and is more formally known as the Lost Mine of Enchantment. This New Mexico mine is near the state's southeastern town of Ruidoso in the Sacramento Mountains. It is well-known as the first new mine discovered in New Mexico since the days of Coronado in the 1500s. Turquoise from Enchantment has the deep green color and beautifully complimentary tan or golden brown matrix typical of high-quality gems. Enchantment turquoise is sometimes found in a range of deep blue colors as well. The green results from the stone's iron content, while the blue comes from the copper content.

Fox (Nevada) mine turquoise holds the distinction of the being the largest turquoise mine in Nevada. Located near Lander County, its turquoise was first mined in prehistoric times, and the mine was known as Cortez. Rediscovered around 1911, the Fox mine has brought forth more than half a million pounds of material. While most recognize the Fox name, it is rightfully called White Horse Fox for the region in Lander County where this high-grade blue gem has been mined.

Emerald Valley turquoise. Pendant. Silver setting. Navajo, signed E. 2¼" x 1¼". *JR's Southwestern Trading Post.*

Fox turquoise. Assorted cut and polished stone. 22 grams. *Courtesy of John Renner. JR's Southwestern Trading Post.*

Fox turquoise. Cuff bracelet. Silver setting. Navajo. 1" x 3". *JR's Southwestern Trading Post.*

Fox turquoise. Cuff bracelet. Multiple stone. Silver setting. Navajo. 2" x 3". *JR's Southwestern Trading Post.*

Fox turquoise. Necklace and pendant. Multiple stones. Silver setting. Navajo, signed Paul Livingston. *Courtesy of John Miller.*

Fox turquoise. Bolo tie. Silver setting with silver tips. Navajo, signed DM. 2¼" x 2". *JR's Southwestern Trading Post.*

Fox turquoise. Cuff bracelet.
114 square-cut cuff. Silver setting.
Navajo, signed Ernest Rangle.
Courtesy of John Miller.

Fox turquoise. Snake-eye cuff bracelet. Silver setting. Unmarked.
Courtesy of John Miller.

Fox (Wild Horse) turquoise. Pendants. Silver setting. Navajo. 1¾" × 3". *JR's Southwestern Trading Post*.

Fox (Wild Horse) turquoise. Cuff bracelet. Silver setting. Navajo, signed Carl Quintano; left, E., middle, right. 1¼" × 1", 2" × 1", 1" × 1". *JR's Southwestern Trading Post.*

Fox (Wild Horse) turquoise. Concho belt. Multiple stones, cluster setting. Signed C. L. 3½" x 2¾". *JR's Southwestern Trading Post*.

Fox turquoise. Pendant and bead necklace. Multiple stones, pueblo-style channel inlay. Silver setting. Kewa Pueblo, signed Clarence Chama. *Courtesy of John Miller.*

Gleeson or Courtland (Arizona) mine turquoise was first extracted in the 1890s. The area was originally settled as a mining camp called Turquoise and had been mined by Native Americans in the area. The Turquoise post office, established in 1890, lasted only until 1894. When John Gleeson registered a copper claim and opened the Copper Belle Mine, the town of Gleeson was created just downhill from the old site of Turquoise in Turquoise Mountain on the south side of the Dragoon Mountains in Cochise County. Little is known of the material mined from the area. Gleeson is better known for its location; take a drive out of this ghost town to the west and you'll run right into Tombstone. Turquoise Mountain is likely the most common name given to a turquoise-bearing hillside, the most popular being the Turquoise Mountain claims outside of Kingman, Arizona.

Godber-Burnham (Nevada) mine turquoise has also been referred to as Dry Creek turquoise. The mine is in northeastern Nevada and this stone is usually white to light blue. Turquoise from this locale is quite hard due to the high levels of aluminum. The material is nearly always found with matrix that is light yellow or brown-gray to gray-black.

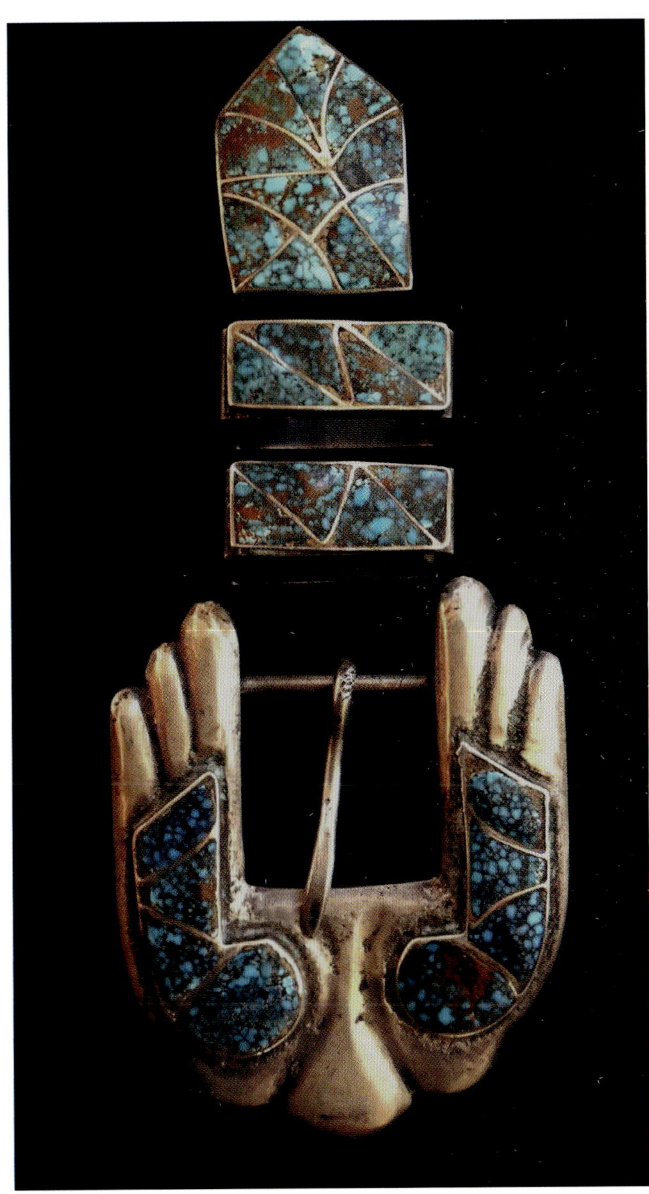

Burnham turquoise. Inlay Ranger Buckle set. Silver setting. Collection of Lynn Trusdell. *Courtesy of John Miller.*

Burnham turquoise. Bracelet. Multiple stones. Bump-out stamp. Silver setting. C. 1930s–1940s. Unmarked. *Courtesy of John Miller.*

Godber and Burnham turquoise. Squash blossom necklace, ring, earring set. Multiple stones. Silver setting. C. 1970s. *Courtesy of John Miller.*

Godber and Burnham turquoise. Pendant. Silver setting. 1¾" × 1½". *JR's Southwestern Trading Post.*

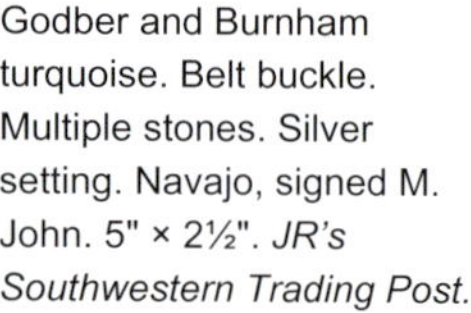

Godber and Burnham turquoise. Belt buckle. Multiple stones. Silver setting. Navajo, signed M. John. 5" × 2½". *JR's Southwestern Trading Post.*

Hachita (New Mexico) mine turquoise has not been extracted since 1905. While it dates to ancient times, it is rare and difficult to acquire. It is generally deep blue and can be confused with the turquoise of Tyrone.

Hachita turquoise. Pin. Silver setting. Navajo, signed Santa Fe DEU. 3" × 1½". *JR's Southwestern Trading Post.*

Hachita turquoise. Belt buckle. Silver setting. Navajo. 3" × 2½". *JR's Southwestern Trading Post.*

Hachita and Chinese turquoise. Belt buckle. Multiple stones, cluster. Silver setting. Navajo, signed DC. 3½" × 2¼". *JR's Southwestern Trading Post.*

Hachita turquoise. Belt buckle. Silver setting. Navajo, signed WBH. 2½" × 2". *JR's Southwestern Trading Post.*

Hachita turquoise. Cuff bracelet. Silver setting. Navajo. 3" × 3½". *JR's Southwestern Trading Post.*

Hachita turquoise. Concho belt. Silver setting. Navajo. 3½" × 2½", 2¼" × 1½", 2¼". *JR's Southwestern Trading Post.*

Hachita turquoise. Concho belt. Multiple stones. Silver setting. Navajo. 3½" × 3", 3" × 2¾". *JR's Southwestern Trading Post.*

Hachita turquoise. Pendant. Multiple stones. Silver setting. Navajo, signed Ray Begay. 3½" × 1¼". *JR's Southwestern Trading Post.*

Halloran Spring (California) mine claims began with the gem in 1896 near Solomons Knob. The first real mention of turquoise made by G. F. Kunz in 1905 was of seams stained with blue. Upon further digging the gem showed a fine blue color. Halloran Spring was worked as various claims, much like others in the Southwest in the late 1800s and early 1900s, and took on the name of the general area rather than the specific mine. There was East Camp, Middle Camp, and West Camp. Much like other claims and mines of the Old West, there is no mention of any further mining operations after 1904. Finding and accurately identifying Halloran Spring turquoise often requires the opinion of an expert in Southwest American turquoise.

Harcross (Nevada) mine turquoise is pure gem turquoise. Founded in 1908 and sold to Walter Godber (of the Godber and Burnham mine), the claim was later assumed by F. B. Cross and J. J. Harrison as the Harcross. Turquoise from this mine is especially hard and ranges in color from blue to green, often with limonite giving it a fine matrix. Though its history is rarely discussed, this material is still worked in small lots by stone cutters who favor its beautiful color.

Indian Mountain turquoise. Cuff bracelet. Multiple stones. Silver setting. Signed. 1" × 3". *JR's Southwestern Trading Post.*

High Lonesome (New Mexico) mine turquoise almost goes too far back to remember. For more than thirty years it is said that the High Lonesome Turquoise Mine owner and crew would look for the recognizable New Mexico green to powder blue turquoise each year over a six-week period. Located in the rugged foothills of southern New Mexico in the lower Hatchet Mountain range, the now obsolete High Lonesome Turquoise Mine is just a memory. Like most other old New Mexico turquoise, it appears High Lonesome yielded little turquoise, so the stone is extremely difficult to find and identify. High Lonesome is known almost exclusively for the name, which is painted on a watering tank and aptly describes the land around it.

Indian Mountain turquoise. Cuff bracelet. Multiple stones. Silver setting. Navajo. 1¾" × 3". *JR's Southwestern Trading Post.*

Indian Mountain (Nevada) mine turquoise was discovered in 1970 by a Native American Shoshone sheep herder. Located in Lander County, Nevada, Indian Mountain is one of the best-known contemporary turquoise mines. The vein of turquoise was subsequently mined by an American family whose marketing efforts were associated with some of the best-known southwestern jewelry artists. The beautiful blue of Indian Mountain lends itself to fine pieces of modern Native American wearable art. In the late 1970s, turquoise from this rich lode was prominently featured in the highly regarded western magazine, *Arizona Highways*. This issue of the state magazine is a prized collectible.

Indian Mountain turquoise. Concho belt. Silver setting. Signed D Gordon. 4½" × 3¼", buckle; 3¼", Concho. *JR's Southwestern Trading Post.*

Indian Mountain turquoise. Cuff bracelet. Silver setting. Signed John Renner. 3¾" × 3". *JR's Southwestern Trading Post.*

Indian Mountain turquoise. Foliage Pendant. Silver setting. C. 1960–1970. *Courtesy of John Miller.*

Ithaca Peak (Arizona) mine is in northwestern Arizona above the Kingman mine, and is a good example of the way turquoise color can vary in one location. While most prized for its striking blue color with pyrite matrix, Ithaca Peak turquoise runs from medium dark green to blue, and is one of the best cut and polished gems in the US. Because primary mining is focused on Kingman and Turquoise Mountain, Ithaca Peak is somewhat harder to find on the market, whether in its natural state or cut and polished.

Kingman turquoise. Natural rough stone. Assorted stones. 220 grams total. *Courtesy of John Renner Collection.*

Johnny Bull (New Mexico) mine is in southeastern New Mexico in Hidalgo County. The Johnny Bull Mine was known to produce light blue to medium blue turquoise bearing brownish matrix and was in production from 1886 to 1917. If you make the trip across the rugged terrain you will be rewarded with specimens that reportedly can still be found, as long as you get proper permission. There are five other abandoned claims in this area.

Kennecott Bingham Canyon (Utah) mine is one of the largest open-pit copper mines in the world and lies about twenty miles southwest of downtown Salt Lake, in the Oquirrh mountains. Kennercott Bingham Canyon is one of the only known turquoise mines in Utah and has produced stones ranging from blue-green to medium blue that take an excellent polish. The turquoise often contains large amounts of pyrite matrix. Given that all resources are being used to extract copper and molybdenum, you rarely see turquoise from this mine today.

Kingman turquoise. Assorted natural rough, cut and polished stones.14 grams total. *Courtesy of John Renner Collection.*

Kingman (Arizona) mine turquoise has become a standard by which other turquoise stones are judged. One of the largest turquoise mines in North America, Kingman has a stature all its own. This stature is reflected in the term Kingman, referring to the intense blue coloring this gemstone most often exhibits. While few mines produced turquoise nuggets, Kingman natural nuggets are famed for their roundness and matrix. One of the most recognized names in turquoise the world over, Kingman is truly a leader in high-quality cut and polished stones.

Kingman turquoise. Polished stone. *JR's Southwestern Trading Post.*

Kingman turquoise. Belt buckle. Silver setting. Signed John Renner. 5" × 4½". *Courtesy of the artist. JR's Southwestern Trading Post.*

Kingman turquoise. Belt buckle. Multiple stones, cluster. Silver setting. Navajo, signed JR. 3¼" × 3". *JR's Southwestern Trading Post.*

Kingman turquoise. Belt buckle. Multiple stones. Silver setting. Signed John Renner. 3" × 5". *Courtesy of the artist.*

Kingman turquoise. Belt buckle. Silver setting. Signed John Renner. 4" × 2½", right; 4" × 6", left. *Courtesy of the artist.*

Kingman turquoise. Belt buckle. Multiple stones, cluster. Silver setting. Navajo, signed Yazzie. 2½" × 3½". *JR's Southwestern Trading Post.*

Kingman turquoise. Belt buckle. Multiple stones, cluster. Silver setting. Navajo, signed Tommy Moore. 3¼" × 4¼". *JR's Southwestern Trading Post.*

Kingman turquoise. Belt buckle. Silver setting. Navajo. 4¼" × 3¾". *JR's Southwestern Trading Post.*

Kingman turquoise. Belt buckle. Silver setting. Navajo. 2½" × 3½". *JR's Southwestern Trading Post.*

Kingman turquoise. Belt buckle. Turquoise, coral, mother-of-pearl, and malachite, Multiple stones, inlay. Navajo. 2" × 3". *JR's Southwestern Trading Post.*

Kingman turquoise. Belt buckle. Silver setting. Signed John Renner. 4½" × 6". *Courtesy of the artist.*

Kingman turquoise. Belt buckle. Multiple stones, inlay. Silver setting. 4¼" × 5¾". *JR's Southwestern Trading Post.*

Kingman turquoise. Necklace. Multiple stones with silver beads. Navajo. 12". *JR's Southwestern Trading Post.*

Kingman turquoise. Belt buckle. Multiple stones, cluster. Silver setting. Navajo. 2½" × 3". *JR's Southwestern Trading Post.*

Kingman turquoise. Belt buckle. Multiple stones. Silver setting. Signed John Renner. 2¼" × 3¼". *Courtesy of the artist.*

Kingman turquoise. Necklace. Multiple stones with silver beads. Pueblo. 28". *JR's Southwestern Trading Post.*

Kingman turquoise. Necklace. Multiple stones with inlay. Pueblo. 22". *JR's Southwestern Trading Post.*

Kingman turquoise. Necklace. Multiple stones with silver beads, Navajo. 21". *JR's Southwestern Trading Post.*

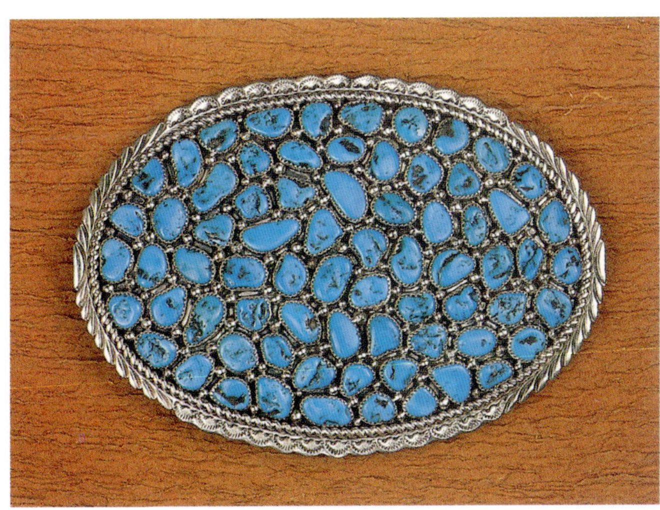

Kingman turquoise. Pendant. Multiple stones, cluster. Silver setting. Navajo, signed V. 4¼" × 6". *JR's Southwestern Trading Post.*

Kingman turquoise. Pendants. Turquoise and coral. Multiple stones. Silver setting. Navajo, signed JR (Jerry Roan). 4" × 6". *JR's Southwestern Trading Post.*

Kingman turquoise. Necklace. Multiple stones. Navajo. 30". *JR's Southwestern Trading Post.*

Kingman and Chinese turquoise. Cross pendants and pin. Turquoise and coral. Multiple stones. Silver setting. Navajo. 1"–2½". *JR's Southwestern Trading Post.*

Kingman turquoise. Necklace. Multiple stone beads with silver. Navajo. 18". *JR's Southwestern Trading Post.*

Kingman turquoise. Pin. Turquoise, coral, shell, onyx, and agate. Multiple stones, inlay. Silver setting. 1" × 3". *JR's Southwestern Trading Post.*

Kingman and Chinese turquoise. Necklaces. Turquoise and coral. Multiple stones, multi-strand. Navajo. 22", 28". *JR's Southwestern Trading Post.*

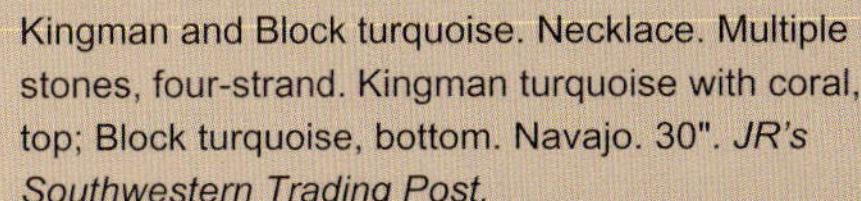

Kingman and Block turquoise. Necklace. Multiple stones, four-strand. Kingman turquoise with coral, top; Block turquoise, bottom. Navajo. 30". *JR's Southwestern Trading Post.*

Kingman turquoise. Pin. Silver setting. Navajo. 2". *JR's Southwestern Trading Post.*

Kingman turquoise. Necklace. Multiple stones, inlay, four-strand beads. Silver setting. Pueblo. 22". *JR's Southwestern Trading Post.*

Kingman turquoise. Pendants. Multiple stones. Silver setting. Navajo. 1" × 3". *JR's Southwestern Trading Post.*

Kingman turquoise. Pendant. Multiple stones. Silver setting. Navajo, signed LLY. 2". *JR's Southwestern Trading Post.*

Kingman turquoise. Fetish necklace. Turquoise and coral. Multiple stones. Pueblo. 20". *JR's Southwestern Trading Post.*

Kingman turquoise. Pendant. Silver setting. Navajo, signed E. 1½". *JR's Southwestern Trading Post.*

Kingman turquoise. Pendant. Silver setting. 2". *JR's Southwestern Trading Post.*

Kingman turquoise. Pendant. Silver setting. Navajo, signed Toadlee. 2¾" × 3". *JR's Southwestern Trading Post.*

Kingman turquoise and coral. Bear claw pendants. Turquoise and coral. Multiple stones. Silver setting. Navajo. 1"–3". *JR's Southwestern Trading Post.*

Kingman turquoise. Pendant. Silver setting. Navajo, signed E. 1½" × 2¾". *JR's Southwestern Trading Post.*

Kingman turquoise. Necklace. Multiple stones. Silver setting. Signed John Renner. 18". *Courtesy of the artist.*

Kingman turquoise. Necklace and earrings. Multiple stones, cluster. Silver setting. Navajo, signed ZB. 20". *JR's Southwestern Trading Post.*

Kingman turquoise. Pendant. Silver setting. Bead necklace. Navajo, signed E. 18". *JR's Southwestern Trading Post.*

Kingman turquoise. Pendants. Turquoise, agate, and onyx. Multiple stones, inlay. Silver setting. Pueblo. 2". *JR's Southwestern Trading Post.*

Kingman turquoise. Pendant. Multiple stones. Silver setting. Signed John Renner. 18". *Courtesy of the artist.*

Kingman turquoise. Pendant. Turquoise and shell. Multiple stones. Cerrillos turquoise bead necklace, multiple stones. Pueblo. 14". *JR's Southwestern Trading Post.*

Kingman turquoise. Pendant. Silver setting. Navajo, signed E. 3¼". *JR's Southwestern Trading Post.*

Kingman turquoise. Pendant. Silver setting. Navajo, signed E. 2". *JR's Southwestern Trading Post.*

Kingman turquoise. Pendant. Multiple stones. Silver setting. Navajo. 3½". *JR's Southwestern Trading Post.*

Kingman turquoise. Pendant. Multiple stones. Silver setting. Signed John Renner. 18". *Courtesy of the artist. JR's Southwestern Trading Post.*

Kingman turquoise. Necklace. Fourteen-strand nugget. Silver setting. *Courtesy of John Miller.*

Kingman turquoise. Pendant. Silver setting. Navajo, signed C. 2¾" × 3¾". *JR's Southwestern Trading Post.*

Kingman turquoise. Necklace. Turquoise, coral, onyx, and mother-of-pearl. Multiple stones, inlay. Silver setting. Zuni. 18". *JR's Southwestern Trading Post.*

Kingman and Chinese turquoise. Pendant. Multiple stones over shell, inlay. Necklace. Chinese turquoise, six strands. Navajo, signed C. 2¼" × 2½"; shell, 18". *JR's Southwestern Trading Post.*

Kingman turquoise. Pendant and earring set. Multiple stones. Silver setting. Signed DB. Pendant: 1¼" × 2¼", earrings: 1¼" × ½". *JR's Southwestern Trading Post.*

Kingman turquoise (reconstituted). Fetish necklace. Multiple stones. Navajo. 18". *JR's Southwestern Trading Post.*

Kingman turquoise (reconstituted). Fetish necklace. Turquoise, agate, mother-of-pearl. Multiple stones. 18". *JR's Southwestern Trading Post.*

Kingman and Chinese turquoise. Pendant: Kingman; necklace: Chinese, four-strand beads. Multiple stones. Navajo. 18". *JR's Southwestern Trading Post.*

Kingman turquoise. Necklace. Multiple stones, silver with silver beads. Navajo. 18". *JR's Southwestern Trading Post.*

Kingman turquoise. Belt buckles. Turquoise and coral chip inlay. Silver settings. Navajo, signed S, (left), Denesto, (middle), HB, (right). 3" × 2". *JR's Southwestern Trading Post.*

Kingman turquoise. Earrings. Multiple stones, cluster. Silver setting. Navajo. 2" × 2¼". *JR's Southwestern Trading Post.*

Kingman turquoise. Belt buckle. Multiple stones. Silver setting. Signed John Renner. 4" × 3½". *Courtesy of the artist.*

Kingman turquoise. Earrings. Multiple stones. Silver setting. Zuni. 2¼". *JR's Southwestern Trading Post.*

Kingman turquoise. Belt buckle. Multiple stones. Silver setting. Signed John Renner. 4" × 3½". *Courtesy of the artist.*

Kingman turquoise. Watch cuff bracelet. Multiple stones. Silver setting. Navajo.1½". *JR's Southwestern Trading Post.*

Kingman turquoise. Earrings. Silver. ⅝". *JR's Southwestern Trading Post.*

Kingman turquoise. Earrings. Multiple stones. Silver setting. Navajo. 1½". *JR's Southwestern Trading Post.*

Kingman turquoise. Earrings. Multiple stones. Silver setting. Navajo. 1¼". *JR's Southwestern Trading Post.*

Kingman turquoise. Earrings. Multiple stones. Silver bear claw setting. Navajo. 2½". *JR's Southwestern Trading Post.*

Kingman turquoise. Earrings. Multiple stones. Silver. Navajo. 1¼". *JR's Southwestern Trading Post.*

Kingman turquoise. Earrings. Silver setting. Navajo, signed GD. ⅞". *JR's Southwestern Trading Post.*

Kingman turquoise. Bolo tie. Silver setting and tips. Signed John Renner. 2¼" × 3". *Courtesy of the artist.*

Kingman turquoise. Bola tie. Turquoise and coral. Silver setting. Signed. *Courtesy of the author.*

Kingman turquoise. Bolo tie. Silver setting. Unmarked. *Courtesy of John Miller.*

Kingman turquoise. Bolo tie. Multiple stones. Silver setting with silver tips. Navajo, signed JR. 6" × 4". *JR's Southwestern Trading Post.*

Kingman turquoise. Watch band. Multiple stones. Silver settings, stretch bands. Navajo. 5". *JR's Southwestern Trading Post.*

Kingman turquoise. Watch bands. Turquoise and coral. Multiple stones, chip inlay. Silver settings, stretch bands. Navajo. 4", 5". *JR's Southwestern Trading Post.*

Kingman turquoise. Watch cuff bracelet. Multiple stones. Silver setting. Navajo, signed PB. 3". *JR's Southwestern Trading Post.*

Side view.

Kingman turquoise. Watch cuff bracelet. Multiple stones. Silver setting. Navajo. 2½". *JR's Southwestern Trading Post.*

Side view.

Kingman turquoise. Cuff bracelet. Multiple stones, cluster. Silver setting. Navajo, signed T. 2½" × 3½". *JR's Southwestern Trading Post.*

Side view.

Kingman turquoise. Watch cuff bracelet. Multiple stones, chip inlay. Silver setting. Navajo, signed T. 2½". *JR's Southwestern Trading Post.*

Kingman turquoise. Cuff bracelet. Multiple stones. Gold setting. Signed John Renner. 2½" × 3". *Courtesy of the artist. Photo: Courtesy of John Renner.*

Kingman turquoise and faceted topaz. Cuff bracelet. Multiple stones. Gold setting. Signed John Renner. 1½" × 3". *Courtesy of the artist. Photo: Courtesy of John Renner.*

Kingman turquoise. Cuff bracelet. Multiple stones, cluster. Silver setting. Navajo, signed T. 4" × 3½". *JR's Southwestern Trading Post.*

Detail.

Kingman turquoise. Bracelet. Turquoise and coral. Silver setting. C. 1970s. Unmarked. *Courtesy of John Miller.*

Kingman turquoise. Bear bracelet. Silver setting. Marked "PAC." *Courtesy of John Miller.*

Kingman turquoise. Bracelet. Nugget cluster old-style. Multiple stones. Silver setting. *Courtesy of John Miller.*

Kingman turquoise. Bracelet. Multiple stones. Silver setting. Navajo (attributed to): Kirk Smith. *Courtesy of John Miller.*

Kingman turquoise. Bracelet. Heavy stamp. Silver setting. Unmarked. *Courtesy of John Miller.*

Kingman turquoise. Cuff bracelet, Multiple stones, cluster. Silver setting. Navajo. 2¼" × 3". *JR's Southwestern Trading Post.*

Kingman turquoise. Cuff bracelet. Turquoise and coral. Multiple stones, chip inlay. Silver setting. Navajo, signed T. 2¼" × 3½". *JR's Southwestern Trading Post.*

Kingman turquoise. Cuff bracelet. Turquoise and coral. Multiple stones, chip inlay. Silver setting. Navajo. 3" × 3". *JR's Southwestern Trading Post.*

Kingman turquoise. Watch cuff bracelet. Turquoise and coral, multiple stones. Silver setting. Navajo. 2½". *JR's Southwestern Trading Post.*

#8, Kingman, and Pilot Mountain turquoise. Rings. Gold and silver settings. *JR's Southwestern Trading Post.*

Kingman turquoise (left and right views). Cuff bracelet. Multiple stones, cluster. Silver setting. Navajo. 3" × 3". *JR's Southwestern Trading Post.*

Kingman turquoise. Cuff bracelet. Multiple stones. Silver setting. Navajo. 2½" × 3". *JR's Southwestern Trading Post.*

Kingman turquoise. Cuff bracelet. Multiple stones, cluster. Silver setting. Navajo. 4" × 3½". *JR's Southwestern Trading Post.*

Kingman turquoise. Cuff bracelet. Multiple stones, cluster. Silver setting. Navajo. 4" × 3½". *JR's Southwestern Trading Post.*

Kingman turquoise. Cuff bracelet. Turquoise and coral. Multiple stones. Silver setting. Navajo. 2½" × 3½". *JR's Southwestern Trading Post.*

Kingman turquoise. Cuff bracelet. Multiple stones. Silver setting. Signed John Renner. 3" × 3½". *Courtesy of the artist.*

Kingman turquoise. Bracelet. Silver setting. Signed John Renner. 4" × 3". *Courtesy of the artist.*

Kingman turquoise. Cuff bracelet. Multiple stones. Silver setting. Signed John Renner. 1½" × 3". *Courtesy of the artist.*

Kingman turquoise. Bracelet. Multiple stones, nugget. Silver setting. Navajo. ⅞" × 3". *JR's Southwestern Trading Post.*

Kingman turquoise. Cuff bracelet. Multiple stones. Silver setting. Signed John Renner. 2" × 3". *Courtesy of the artist.*

Kingman turquoise. Cuff bracelet. Multiple stones, nugget. Silver setting. Signed John Renner. 1½" × 3". *Courtesy of the artist.*

Kingman turquoise. Bracelet. Multiple stones. Silver setting. Signed John Renner. 2¼" × 3". *Courtesy of the artist.*

Kingman turquoise. Cuff bracelet. Multiple stones. Silver setting. Navajo. 2½" × 3". *JR's Southwestern Trading Post.*

Kingman turquoise. Assorted animal carvings. Turquoise, agate, and coral. 1"–4". *JR's Southwestern Trading Post.*

Kingman turquoise. Concho belt. Multiple stones. Silver setting. Signed John Renner. 4¾" × 4¼," buckle; 4½" × 4¾", Concho. *Courtesy of the artist.*

Detail.

Kingman turquoise. Concho belt. Multiple stones. Silver setting, leather. 2¼" × 50". *JR's Southwestern Trading Post.*

Kingman turquoise. Concho belt. Multiple stones, cluster. Silver setting, leather. 2¼" × 50". *JR's Southwestern Trading Post.*

Kingman turquoise. Belt. Multiple stones. Silver setting, leather. Signed John Renner. 2" × 50". *JR's Southwestern Trading Post.*

Kingman and Chinese turquoise, coral, lapis. Rings. Silver settings. Kingman Turquoise, coral, and lapis, left; Kingman Turquoise, coral, and lapis, middle; Chinese Turquoise, right. *JR's Southwestern Trading Post.*

Kingman turquoise. Necklace. Multiple stones, natural rough. Gold setting. Signed John Renner. 1¾" × 1¼" stones. *Courtesy of the artist.*

Landers, Sleeping Beauty, and Kingman turquoise. Rings. Silver settings. *JR's Southwestern Trading Post.*

Landers turquoise. Cuff bracelet. Silver setting. Navajo. 2½" × 3". *JR's Southwestern Trading Post.*

Lander Blue (Nevada) mine turquoise has a unique discovery story that adds to its value. In 1973, Rita Hapgold, a Nevada blackjack dealer, collected some turquoise nuggets while picnicking at Indian Creek. She went on to claim the site, naming it the Mary Louise Lode Mining Claim. Later that year she sold her claim to two others who established the Lander Blue Turquoise Company. Pound for pound, Lander Blue has become one of the most valuable turquoises in the world. Referred to by many as a "hat mine" because a hat could cover the recovery area, Lander Blue gave up a meager ninety-eight pounds of turquoise, about two buckets full. Rich in contemporary history and exquisite in its dark blue color with dark matrix, Lander Blue is a Southwest classic.

Leadville (Colorado) is one of the few and arguably best of Colorado's turquoise mining districts. The Sugarloaf District in Lake County, Colorado, includes a number of turquoise deposits sold under the name Leadville. Sugarloaf District claims such as the Josie May, the Turquoise Chief, and the Poor Boy are examples of Leadville area mine names. Due to the strict mining and environmental laws in Colorado today, little work is being done on any claims. The Turquoise Chief Turquoise Mine, also called the Josie May Turquoise Mine, is the most well-known turquoise mine in the Leadville district. The beautiful, vibrant-color gemstones exhibit a unique appearance and are thus somewhat more easily identified with the Leadville region. Little material has been seen from this locality in many years. High-grade Leadville turquoise has a lively color, exceptional hardness, and interesting matrix patterns.

New Landers turquoise. Matching cuff bracelet and ring. Silver setting. Navajo, signed Tommy Jackson. *Photo courtesy of Western Trading Post.*

Lone Mountain (Nevada) mine, also known as Blue Jay Mine, is known for maintaining its color over time. Located in Esmeralda County, the turquoise is often found in nodules and has a fairly wide blue color range, some including spider webbing. Lapidaries and jewelers have a particular fondness for Lone Mountain because it holds its color well and therefore is used to create spectacular jewelry.

Lone Mountain turquoise. Ring. Gold setting. 1½" × ¾". *JR's Southwestern Trading Post.*

Lone Mountain turquoise. Cuff bracelet. Silver setting. Navajo, signed N. 1½" × 3". *JR's Southwestern Trading Post.*

Lone Mountain, Red Mountain, Fox, and Hachita turquoise. Rings. Gold and silver settings. *JR's Southwestern Trading Post.*

Lone Mountain, Chinese, and Morenci turquoise. Rings. Gold and silver settings. *JR's Southwestern Trading Post.*

Lone Mountain turquoise. Ring. Silver setting. Navajo, signed Darrell Dean Begay (DDB). *Courtesy of John Miller.*

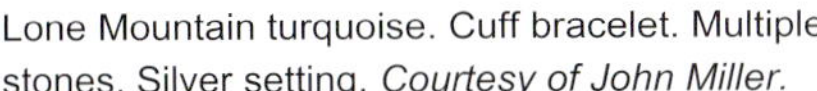

Lone Mountain turquoise. Cuff bracelet. Multiple stones. Silver setting. *Courtesy of John Miller.*

Manassa (Colorado) mine turquoise can be found in south-central Colorado. One of the few notable Colorado mines, Manassa, also known as King's Manassa Mine, named for the family currently working the rock, has revealed turquoise of blue-green to green with a lovely yellow to brown matrix. This truly unique turquoise gem is still in production.

Manassa turquoise. Cuff bracelet. Silver setting. Navajo. 1¾" × 3". *JR's Southwestern Trading Post.*

Manassa turquoise. Bolo tie. Silver setting with silver tips. Navajo. 4½" × 3". *JR's Southwestern Trading Post.*

Mannassa turquoise. Navajo squash blossom. Multiple stones. Silver setting. Navajo, signed Douglas Etsitty. C. 1980. Verified by the grandson of the mine owner (who likely also cut the stones). *Photo courtesy of Western Trading Post.*

Mojave turquoise. Belt buckle. Multiple stones. Silver setting. Signed John Renner. 5" × 4". *Courtesy of the artist.*

Mojave turquoise. Belt buckle. Multiple stones. Silver setting. Signed John Renner. 5" × 4½". *Courtesy of the artist.*

Mojave (Montana) mine turquoise is a beautiful and difficult-to-acquire green turquoise often referred to as Green Apple turquoise. It is good quality and takes a nice polish. It is exceptional when set in silver and is truly one of the more unusual turquoise materials for its intense and consistent coloring.

Morenci (Arizona) mine turquoise is one of the best-known turquoise names in the world. Ranking in recognition with Kingman and Bisbee, also Arizona mines, Morenci turquoise is quite distinguishable for its pyrite inclusions that when polished take on a silver sheen. Most Morenci turquoise runs from a beautiful dark sky blue to light blue. Many know of Morenci turquoise because it was the first American mined turquoise to come to market. Worked by many craftsmen into Native American jewelry, Morenci can be found in some stunning older pieces, and though the mine is now depleted, you can still find contemporary artisans working the material, making it highly collectible.

Morenci turquoise. Assorted natural blue-green stones. 84 grams total. *Courtesy of John Renner Collection.*

Morenci turquoise. Assorted natural #1 green stones. 60 grams total. *Courtesy of John Renner Collection.*

Morenci turquoise. Pendant. Multiple stones. Silver setting. Navajo. 3" × 2¼". *JR's Southwestern Trading Post.*

Morenci turquoise. Necklace. Turquoise and coral. Multiple stones, single strand coral beads. Pueblo. 28". *JR's Southwestern Trading Post.*

Morenci turquoise. Squash blossom necklace. Multiple stones. Silver setting. Navajo, signed Rajh. 26". *JR's Southwestern Trading Post.*

Morenci turquoise. Belt buckle. Silver setting. Navajo, signed RNS. 3" × 2". *JR's Southwestern Trading Post.*

Morenci turquoise. Earrings. Silver setting. Navajo. 1" × $1\frac{1}{8}$". *JR's Southwestern Trading Post.*

Morenci turquoise. Watch band. Multiple stones. Silver setting, stretch band. Navajo. 5". *JR's Southwestern Trading Post.*

Morenci turquoise. Bracelet. Multiple stones. Silver setting. Navajo, signed MA. 1½" × 3". *JR's Southwestern Trading Post.*

Morenci turquoise. Cuff bracelet. Turquoise and coral. Silver setting. Navajo, signed J. W. C. 2" × 3½". *JR's Southwestern Trading Post.*

Morenci turquoise. Cuff bracelet. Silver setting. Navajo, signed Tom. 2½" × 3". *JR's Southwestern Trading Post.*

Morenci turquoise. Bracelet. Note the "organic" foliate (leaf) carries through into the pyrite. Multiple stones. Silver setting. Navajo, signed David Zachary. *Courtesy of John Miller.*

Morenci and Sleeping Beauty turquoise. Rings. Silver settings. *JR's Southwestern Trading Post.*

Morenci turquoise. Bolo tie. Silver setting. Hopi, signed Alan Pooyouma. *Courtesy of John Miller.*

Morenci turquoise. Bracelet. Channel inlay, multiple stones. Silver setting. C. 1950–1960. *Courtesy of John Miller.*

Morenci turquoise. Bracelet and pendant necklace. Multiple stones. Silver setting. Navajo, signed Emma Lincoln. Unmarked. *Courtesy of John Miller.*

Morenci turquoise. Bracelet. Multiple stones. Silver setting. Marked "C." *Courtesy of John Miller.*

Morenci turquoise. Bracelet. Multiple stones. Silver setting. Navajo, signed Emma Lincoln. Unmarked. *Courtesy of John Miller.*

Morenci turquoise. Squash Blossom necklace, belt, and bracelet. 441 individual "Seafoam" nuggets. Multiple stones. Silver setting. Navajo, signed Mark Chee. Santa Fe abstract-traditionalist artistic influence in the 1940s–1960s. World War II Veteran, Chee is buried in the National Cemetery, Santa Fe. *Courtesy of John Miller.*

Morenci turquoise. Bracelet with pyrite. "Lady Sturgis." Silver setting. Unmarked. *Courtesy of John Miller.*

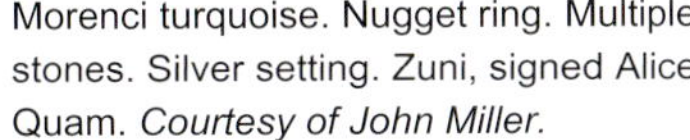

Morenci turquoise. Nugget ring. Multiple stones. Silver setting. Zuni, signed Alice Quam. *Courtesy of John Miller.*

Nevada Blue (Nevada) mine turquoise is some of the finest turquoise mined in Nevada. Nevada Blue has also been known as the Pinto or Watts mine. Near the crest of the Shoshone Range in Lander County, the Nevada Blue was first discovered in 1901 by Jim Watt. Southwest mines are characteristically difficult to access, and the Nevada Blue is no exception. High-grade Nevada Blue ranges in color from medium to dark blue and typically has distinguishable brown or black spider web matrix. Throughout the 1970s Nevada Blue turquoise was well marketed as a turquoise of choice worked by many of the Southwest's most creative and talented silversmiths. The April 1979 issue of *Arizona Highways* featured museum-quality Nevada Blue turquoise set in the highest quality jewelry.

Number (#) 8 (Nevada) mine produces turquoise with character. Few gemstones exhibit such variety in appearance, making it one of the easier turquoises to identify. This gemstone is a bright powder blue to green with golden brown to black spider web matrix that creates distinctive patterns. The Eureka County mine was discovered in 1925 and first mined in 1929; no new material has been mined since 1976. The last mine owner, Dowell Ward, stockpiled material for later sorting, so Number (#) 8 turquoise continued to find its way to market, where it enhanced some of the finest contemporary Southwest silversmith pieces ever created. The unsorted stock has now been depleted, dispersed to gem cutters and collectors. The Gold Mining Company now owns the claim, and the Number (#) 8 mine has become part of its gold mining operations.

#8 turquoise. Assorted cut and polished stone. 20 grams. *Courtesy of John Renner.*

#8 turquoise. Polished slab and cabochons. *Photo courtesy of Western Trading Post.*

#8 turquoise. Cabochons. *Photo courtesy of Western Trading Post.*

#8 turquoise. Cuff bracelet. Multiple stones. Silver setting. Signed John Renner. 1½" × 3". *JR's Southwestern Trading Post.*

#8 turquoise. Belts, bracelet, necklace. Multiple stones. Silver setting. Zuni, signed Warren (Doris) Ondelacy. Personal price code by Zuni trader: C.G. Wallace. *Courtesy of John Miller.*

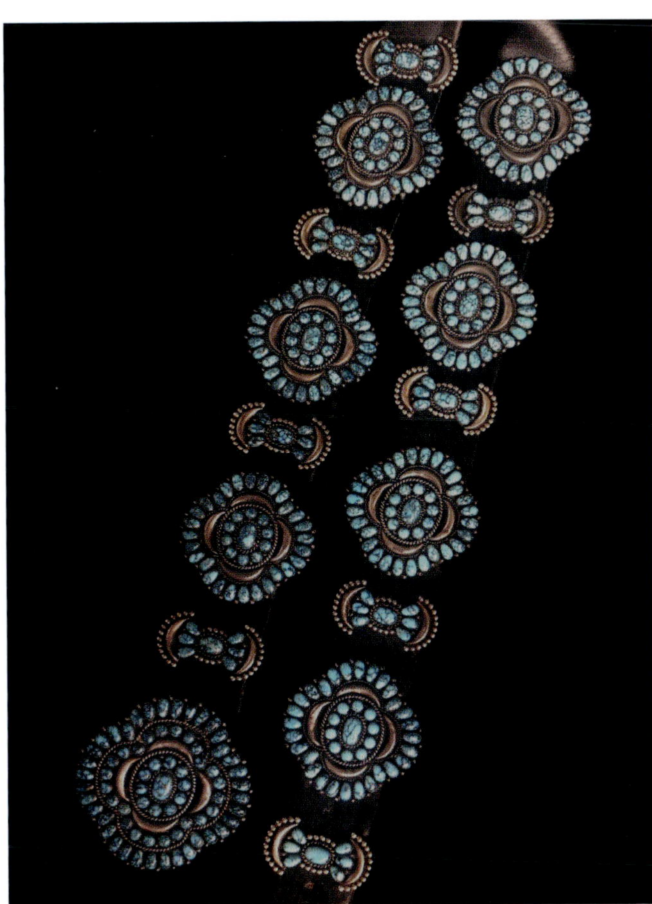

#8 turquoise. Belt, buckle. Multiple stones. Silver setting. C. early 1950s. Zuni, signed Warren Ondelacy. C. G. Wallace price code inscribed into underside of buckle. *Courtesy of John Miller.*

#8 turquoise. Belt, buckle, and first butterfly spacer. Multiple stones. Silver setting. C. early 1950s. Zuni: Warren Ondelacy. C. G. Wallace price code inscribed into underside of buckle. *Courtesy of John Miller.*

#8 turquoise. Bracelet. Hammered wire. Multiple stones. Silver setting. C. 1940s. Marked UITA (United Indian Trader's Association, an industry "consumer protection" marking certifying Native authenticity). *Courtesy of John Miller.*

#8 turquoise. Cuff bracelet. Silver setting. Navajo. 2" × 3". *JR's Southwestern Trading Post.*

#8 turquoise. Belt buckle. Multiple stones. Silver setting. Navajo. 2½" × 2¾", buckle. *JR's Southwestern Trading Post.*

#8 turquoise. Cuff bracelet. Multiple stones. Silver setting. Navajo. 3½" × 3". *JR's Southwestern Trading Post.*

#8 turquoise. Cuff bracelet. Gold and silver setting. Apache, signed Marc Antia. 1½" × 3". *JR's Southwestern Trading Post.*

#8 turquoise. Cuff bracelet. Silver setting. Signed John Renner. 3" × 3". *JR's Southwestern Trading Post.*

#8 turquoise. Teardrop earrings. Silver setting. *Courtesy of John Miller.*

#8 turquoise. Bracelet. Silver setting. Navajo, signed Elle C. Jackson. *Photo courtesy of Western Trading Post.*

#8 turquoise. Pendant. Silver setting. Navajo. 2" × 1¾".
JR's Southwestern Trading Post.

#8 turquoise. Cuff bracelet. Silver setting. Anglo, David Zachary.
Courtesy of John Miller.

#8 turquoise. Pendant. Silver setting. Navajo, signed E.
2" × 1½". *JR's Southwestern Trading Post.*

#8 turquoise. Pendant. Silver setting. Navajo, signed E. 3¼" × 1¼".
JR's Southwestern Trading Post.

#8 turquoise. Cuff bracelet. Heavy chisel-file stamp. Silver setting. Navajo, Terry Martinez. *Courtesy of John Miller.*

#8 turquoise. Earrings. Butterfly clip-on. Silver setting. C. 1960s. Marked J. Platero. *Courtesy of John Miller.*

#8 turquoise. Cuff bracelet. Silver setting. Titled: "Blue Corn," collaboration by Navajo tufa cast artist Philander Begay and Kevin Barnhill. *Courtesy of John Miller.*

#8 turquoise. Ring. Deep chisel-stamp. Silver setting. Navajo, Gary Reeves. Tobe Turbin collection. *Courtesy of John Miller.*

#8 turquoise. Squash blossom necklace. Multiple stones. Silver setting. *Courtesy of John Miller.*

#8 turquoise. Necklace. Multiple stones. Silver setting. Navajo, Gary Reeves (deceased, brother of Sunshine Reeves). *Courtesy of John Miller.*

#8 turquoise. Cuff bracelet. Silver setting. C. 1970s. Artist mark "JJM". *Courtesy of John Miller.*

#8 turquoise. Bracelet. Cuttlebone cast. Silver setting. Navajo, Ryan Bennally. *Courtesy of John Miller.*

#8 turquoise. Bracelet. Silver setting. Marked, "MM" (Mary Morgan). *Courtesy of John Miller.*

#8 turquoise. Bracelet. Seamless inlay, multiple stones. Silver setting. Marked, A. Francisco. *Courtesy of John Miller.*

#8 turquoise. Concho Belt. Multiple stones. Silver setting. C. early 1950s. Zuni, signed Warren Ondelacy. C.G. Wallace price code inscribed into underside of buckle. *Courtesy of John Miller.*

Orvil Jack (Nevada) mine is a Nevada mine named for its discoverer, Orvil Jack. Located in the Blue Ridge of the Crescent Valley, Orvil Jack continues to be operated by Jack's daughter. While little material is being extracted, the turquoise is highly sought after. Orvil Jack turquoise has the rare yellow-green coloring that comes from its zinc content. If you find yourself in northern Nevada, consider stopping by to see the mine. Be sure to ask permission, as, like most mines, even those depleted, it is on private property.

Orvil Jack turquoise. Cuff bracelet. Multiple stones. Silver setting. Navajo. 1½" × 3". *JR's Southwestern Trading Post.*

Orvil Jack and Chinese turquoise. Cuff bracelet. Multiple stones. Silver settings. Chinese turquoise, left; Orvil Jack turquoise, right. Signed Tom Dewit, left; Navajo, right. 1" × 3", 1¼" × 3". *JR's Southwestern Trading Post.*

Paiute (Nevada) mine turquoise is in central Nevada, sharing a mountain with the Godber and Burnham mines. In active production only since 1992, the Paiute is one of the region's currently active mines. Though claims at the site date back to the mid-1970s, the limited material coming from Paiute makes it quite desirable. Most of the turquoise is a high-grade material in shades of blue with spider webbing that ranges from black to brown and red to orange. A hard stone, Paiute is readily cut and polished.

Persian (Iran) turquoise deposits have been mined for more than 2,000 years. South of the Caspian Sea, and therefore close to trade routes, the turquoise remains the finest in the world. Over the course of thousands of years, the intense blue Persian turquoise has been synonymous with quality. Today, the term "Persian turquoise" is a definition of quality more than a statement of origin. Its beautiful, variant sky blue color is almost always without matrix and what many think of as "true" turquoise. While the region's mines produce stones in a variety of blue colors, Persian turquoise is one of the easier localities to identify because of its distinctive color and excellence.

Persian turquoise. Assorted cut and polished stones. 28 grams. *Courtesy of John Renner.*

Persian turquoise. Earrings. Multiple stones, cluster. Silver setting. Zuni, signed Wayne Carla. 3½" × 1¼". *JR's Southwestern Trading Post.*

Persian turquoise. Cuff bracelet. Silver setting. 2" × 3". *JR's Southwestern Trading Post.*

Persian turquoise. Watch cuff bracelet. Silver setting. Navajo.1¼" × 3". *JR's Southwestern Trading Post.*

Persian turquoise. Bolo tie. Turquoise and coral. Silver setting with silver tips. Navajo. 3¼" × 2¼". *JR's Southwestern Trading Post.*

Persian turquoise. Beads. Graduated strand, C. late-1960s–1970s. *Courtesy of John Miller.*

Pilot Mountain (Nevada) mine turquoise, also known as the Montezuma or Troy Springs, is at the southern end of the Pilot Mountains in Mineral County. Discovered in 1905 by William Miller of Tonopah, Nevada, the Pilot Mountain mine was originally a small tunnel mine; however, production greatly increased in the 1970s when heavy machinery was used. In recent years the mine has been returned to a small, one-family operation. Primarily found in thin seams, though some larger nuggets have been mined as well, Pilot Mountain is a beautiful, hard turquoise with colors ranging from bright blue, to dark blue with a greenish cast, to very dark blue, all taking an excellent polish. This turquoise often shows blue and green transitioning across the same stone. This unique quality draws collectors and gem cutters to Pilot Mountain turquoise. It is common to find dark brown limonite blotchy patterns associated with this material, some with spider webbing.

Red Mountain (Nevada) mine is one of several turquoise mines in Lander County. Having produced a large quantity of gem-grade material, Red Mountain turquoise is the rival of some of the finest turquoise stones of the American Southwest. Often used by today's leading Native American artisans, the red spider webbing gives this turquoise its distinctive name and a place in any collection.

Red Mountain turquoise. Assorted cut and polished stones. 6 grams. *Courtesy of John Renner.*

Red Mountain turquoise. Necklace. Turquoise and coral. Multiple stones with silver beads. Navajo. 26". *JR's Southwestern Trading Post.*

Pilot Mountain turquoise. Bracelet. Silver setting. C. 1970s. Unmarked. *Courtesy of John Miller.*

Pilot Mountain turquoise. Bracelet. Hopi style silver overlay. Silver setting. Marked, BB. *Courtesy of John Miller.*

Royston (Nevada) has three mines: Bunker Hill, Oscar Wehrend, and the most notable, Royal Blue. Royston district turquoise still produces gem-grade materials that range from dark green to a beautiful light blue, often with brown matrix. This relatively soft turquoise is highly collectible, as the mine produces less workable material than mines producing harder stone.

Royston turquoise. Cuff bracelet. Multiple stones. Silver setting. Navajo, signed C MTZ. 1½" × 3". *JR's Southwestern Trading Post.*

Royston turquoise. Necklace. Multiple stones with shell (Wampum) beads. Pueblo. 26". *JR's Southwestern Trading Post.*

Royston turquoise. Bolo, ring, and bracelet. Turquoise and coral. Silver setting. Known for their incredible micro-inlay, Carl and Anita Clark started by creating traditional stone settings. C. 1973. *Photo courtesy of Western Trading Post.*

Royston turquoise. Pendant. Silver setting. Navajo, signed DH. 3½" × 1¾". *JR's Southwestern Trading Post.*

Santa Rita (New Mexico) mine turquoise is similar to Tyrone turquoise, with a beautiful blue and green color, and some with high-grade gold ore matrix. The similarities with Tyrone turquoise may be due to the mines' close proximity to one another. Santa Rita turquoise is truly top grade; some looks quite similar to the finest old Morenci. Artisans favor Santa Rita turquoise for setting in silver and gold jewelry.

Sleeping Beauty (Arizona) mine yields a favorite turquoise of the Native American Zuni Pueblo Native Americans. Known for its even blue color devoid of matrix, Sleeping Beauty turquoise is frequently found in petit point and inlay jewelry. Collectors can acquire excellent examples of Sleeping Beauty turquoise because it is one of the largest North American turquoise mine operations still actively being worked.

Sleeping Beauty turquoise. Assorted rough stones. 186 grams total. *Courtesy of John Renner Collection.*

Sleeping Beauty turquoise. Belt buckle. Multiple stones, cluster. Silver setting. Signed John Renner. 3" × 2". *Courtesy of the artist.*

Sleeping Beauty turquoise. Assorted cut and polished stones. 10 grams total. *Courtesy of John Renner Collection.*

Sleeping Beauty turquoise. Belt buckle. Turquoise, shell, coral, and black onyx. Multiple stones, inlay. Silver setting. Zuni, signed Russell Sam. 3" × 2". *JR's Southwestern Trading Post.*

Sleeping Beauty turquoise. Belt buckle. Turquoise, shell, and black onyx. Multiple stones, inlay. Silver setting. Zuni, signed TJ. 2¾" × 1¾". *JR's Southwestern Trading Post.*

Sleeping Beauty turquoise. Belt buckle. Turquoise, coral, shell, and malachite. Multiple stones, inlay. Silver setting. Zuni, signed Sammy and Esther Anardian. 2½" × 2". *JR's Southwestern Trading Post.*

Sleeping Beauty turquoise. Belt buckle. Turquoise, agate, and mother-of-pearl. Multiple stones, inlay. Silver setting. Zuni, signed W. 3½" × 2½". *JR's Southwestern Trading Post.*

Sleeping Beauty turquoise. Belt buckle. Turquoise, shell, and black onyx. Multiple stones, inlay. Silver setting. Navajo, signed TJ. 3" × 2". *JR's Southwestern Trading Post.*

Sleeping Beauty turquoise. Belt buckle. Multiple stones. Silver setting. Navajo. 3½" × 3". *JR's Southwestern Trading Post.*

Sleeping Beauty turquoise. Belt buckle. Turquoise, shell, coral, and black onyx. Multiple stones, inlay. Silver setting. Zuni, signed Leander and Lisa Othole. 3" × 2". *JR's Southwestern Trading Post.*

Sleeping Beauty turquoise. Belt buckle. Turquoise, shell, and coral. Multiple stones, cluster. Silver setting. Zuni. 3" × 2½". *JR's Southwestern Trading Post.*

Sleeping Beauty turquoise. Belt buckle. Turquoise, shell, coral, and black onyx. Multiple stones, inlay. Silver setting. Zuni, signed RIH. 3½" × 3". *JR's Southwestern Trading Post.*

Sleeping Beauty turquoise. Belt buckle. Multiple stones, inlay. Silver setting. Navajo. 3½" × 3". *JR's Southwestern Trading Post.*

Sleeping Beauty turquoise. Cross pendant. Multiple stones, inlay. Silver setting. 4" × 2". *JR's Southwestern Trading Post.*

Sleeping Beauty turquoise. Pendant and pins. Multiple stone, petit point. Silver setting. Zuni, signed Penteah. 3", 2½", 1". *JR's Southwestern Trading Post.*

Sleeping Beauty turquoise. Pin. Turquoise, coral, shell, and black onyx. Multiple stones, inlay. Silver setting. Zuni, signed G. Smith and Ray Watson. 1¼" × 1". *JR's Southwestern Trading Post.*

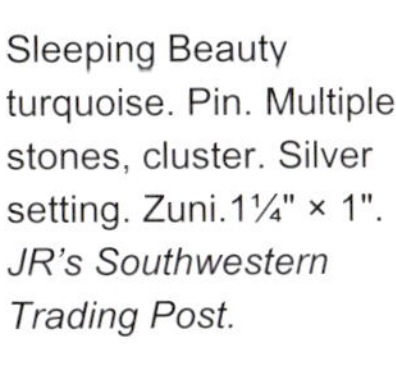

Sleeping Beauty turquoise. Pin. Multiple stones, cluster. Silver setting. Zuni.1¼" × 1". *JR's Southwestern Trading Post.*

Sleeping Beauty turquoise. Pin. Multiple stones. Silver setting. Navajo, signed JR. 3", 2¼". *JR's Southwestern Trading Post.*

Sleeping Beauty turquoise. Pin. Multiple stone, squash blossom. Silver setting. Navajo. 2" × 1". *JR's Southwestern Trading Post.*

Sleeping Beauty turquoise. Pin. Multiple stones, cluster. Silver setting. Navajo. 6" × 4". *JR's Southwestern Trading Post.*

Sleeping Beauty turquoise. Pin. Multiple stones, cluster. Silver setting. Zuni. 3½". *JR's Southwestern Trading Post.*

Sleeping Beauty turquoise. Pendant. Multiple stones. Silver setting. Zuni, signed Effie C. 2½" × 1½". *JR's Southwestern Trading Post.*

Sleeping Beauty turquoise. Pendant. Turquoise and coral. Multiple stones. Silver setting. Zuni, signed Effie C. 2" × 1". *JR's Southwestern Trading Post.*

Sleeping Beauty turquoise. Necklace. Multiple stones, inlay. Silver beads. Silver setting. Navajo. 18". *JR's Southwestern Trading Post.*

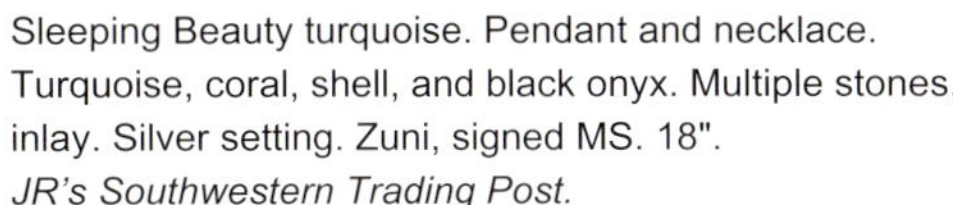

Sleeping Beauty turquoise. Pendant and necklace. Turquoise, coral, shell, and black onyx. Multiple stones, inlay. Silver setting. Zuni, signed MS. 18". *JR's Southwestern Trading Post.*

Sleeping Beauty turquoise. Necklace. Multiple stones, petit point. Silver setting. Zuni. 20". *JR's Southwestern Trading Post.*

Sleeping Beauty turquoise. Necklace. Multiple stones, petit point. Silver setting. Zuni. 18". *JR's Southwestern Trading Post.*

Sleeping Beauty turquoise. Necklace. Turquoise, shell, and coral. Multiple stones. Zuni. 24". *JR's Southwestern Trading Post.*

Sleeping Beauty turquoise. Belt buckle. Multiple stones, cluster. Silver setting. Navajo. 3½" × 3". *JR's Southwestern Trading Post.*

Sleeping Beauty turquoise. Earrings. Multiple stones. Silver setting. Zuni. 1". *JR's Southwestern Trading Post.*

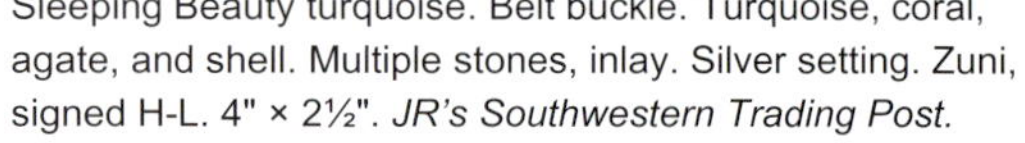

Sleeping Beauty turquoise. Belt buckle. Turquoise, coral, agate, and shell. Multiple stones, inlay. Silver setting. Zuni, signed H-L. 4" × 2½". *JR's Southwestern Trading Post.*

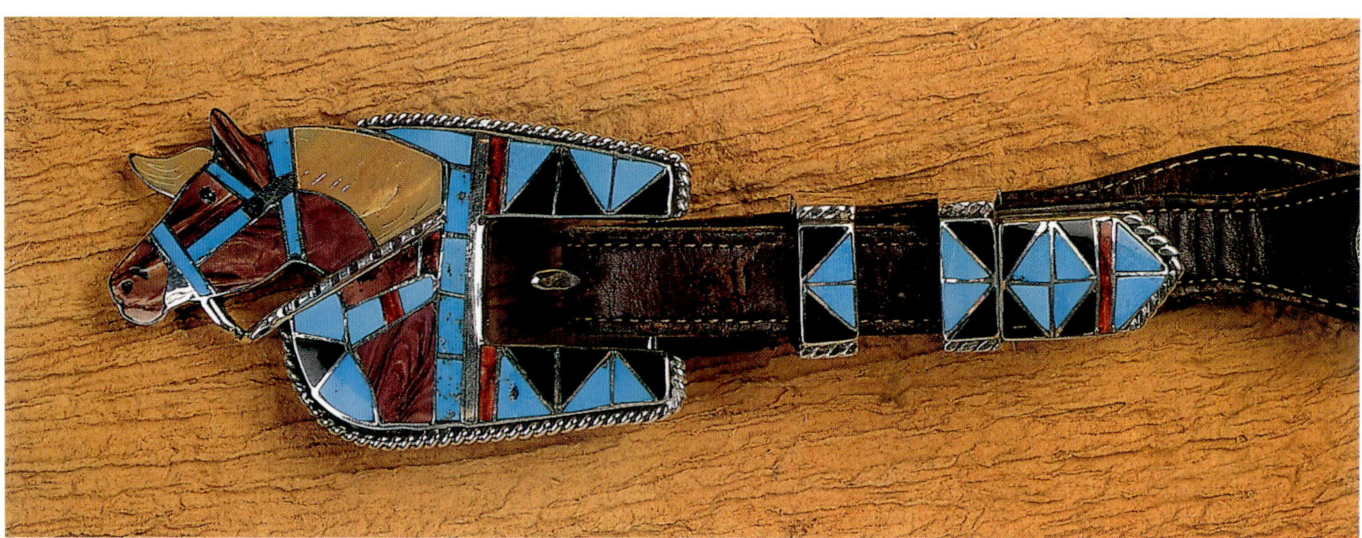

Sleeping Beauty turquoise. Belt buckle. Multiple stones. Silver setting. Zuni, signed Effie C. 3½" × 3". *JR's Southwestern Trading Post.*

Sleeping Beauty turquoise. Earrings. Turquoise, coral, shell, and black onyx. Multiple stones, inlay. Silver setting. Zuni. 1" × 1". *JR's Southwestern Trading Post.*

Sleeping Beauty turquoise. Earrings. Multiple stones. Silver setting. Zuni, signed Effie C. 1¾" × ¾". *JR's Southwestern Trading Post.*

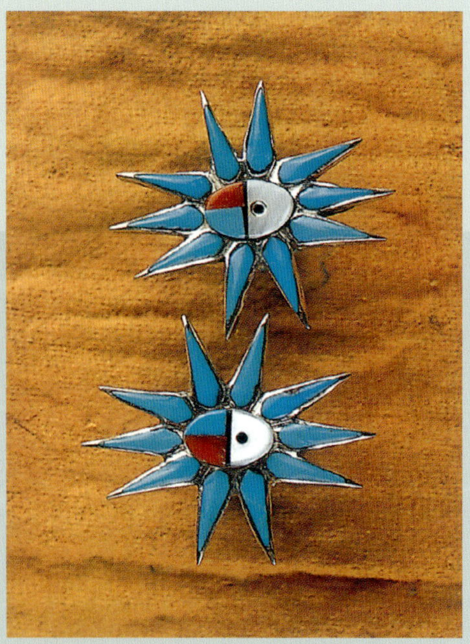

Sleeping Beauty turquoise. Earrings. Turquoise, coral, shell, and black onyx. Multiple stones, inlay. Silver setting. Zuni. 1" × 1". *JR's Southwestern Trading Post.*

Sleeping Beauty turquoise. Earrings. Multiple stones. Silver setting. Zuni, signed Effie C. 1" × ½". *JR's Southwestern Trading Post.*

Sleeping Beauty turquoise. Earrings. Turquoise, coral, shell, and black onyx. Multiple stones, inlay. Silver setting. Navajo. ¾" × ½". *JR's Southwestern Trading Post.*

Sleeping Beauty turquoise. Earrings. Multiple stones, inlay. Silver setting. Navajo. 1". *JR's Southwestern Trading Post.*

Sleeping Beauty turquoise. Bolo tie. Turquoise, coral, shell, black onyx, and sugilite. Multiple stones, inlay and silver tips. Silver setting. Zuni.1¾" × 1¼". *JR's Southwestern Trading Post.*

Sleeping Beauty turquoise. Bolo tie. Turquoise, coral, shell, and black onyx. Multiple stones, inlay and silver tips. Silver setting. Zuni, signed Dev Eriacho. 2¼" × 2". *JR's Southwestern Trading Post.*

Sleeping Beauty turquoise. Bolo tie. Turquoise, coral, shell, and black onyx. Multiple stones, inlay and silver tips. Silver setting. Zuni. 2" × 2¼". *JR's Southwestern Trading Post.*

Sleeping Beauty turquoise. Bolo tie. Turquoise, coral, and lapis lazuli. Multiple stones, inlay and silver tips. Silver setting. Navajo. 2½" × 2½". *JR's Southwestern Trading Post.*

Sleeping Beauty turquoise. Bolo tie. Turquoise, coral, shell, lapis, and black onyx. Multiple stones, inlay and silver tips. Silver setting. Zuni. 6" × 4". *JR's Southwestern Trading Post.*

Sleeping Beauty turquoise. Bolo tie. Turquoise, coral, shell, and black onyx. Multiple stones, inlay. Silver setting. Zuni, signed Roger Cellicion. 6" × 4". *JR's Southwestern Trading Post.*

Sleeping Beauty turquoise. Cuff bracelet. Multiple stones, cluster. Silver setting. Navajo, signed FM Begay. 4" × 3". *JR's Southwestern Trading Post.*

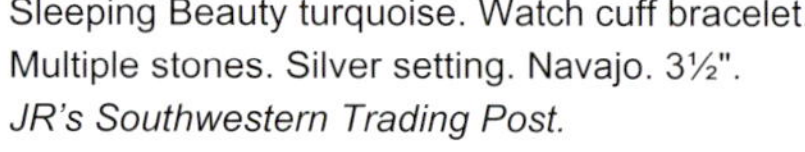

Sleeping Beauty turquoise. Watch cuff bracelet. Multiple stones. Silver setting. Navajo. 3½". *JR's Southwestern Trading Post.*

Sleeping Beauty turquoise. Cuff bracelet. Silver setting. Navajo, signed Roger Nelson. 3½" × 1¾". *JR's Southwestern Trading Post.*

Sleeping Beauty turquoise. Cuff bracelet. Turquoise and coral. Multiple stones. Silver setting. Zuni, signed Effie C. 3" × 1". *JR's Southwestern Trading Post.*

Sleeping Beauty turquoise. Cuff bracelet. Multiple stones. Silver setting. Navajo, signed RC. 3" × 3". *JR's Southwestern Trading Post.*

Sleeping Beauty turquoise. Cuff bracelet. Multiple stones, inlay. Silver setting. Navajo. 3" × 1¼". *JR's Southwestern Trading Post.*

Sleeping Beauty turquoise. Cuff bracelet. Multiple stones. Silver setting. Navajo. 4" × 3". *JR's Southwestern Trading Post.*

Sleeping Beauty turquoise. Concho belt. Multiple stones, cluster buckle. Silver setting. Navajo. 3½" × 3½, 3" × 2¾", 2" × 1¾". *JR's Southwestern Trading Post.*

Sleeping Beauty turquoise. Bracelet. Turquoise and lapis lazuli. Lizard design inlay. Silver setting. Zuni, signed Rick Johns. *Courtesy of John Miller.*

Sleeping Beauty turquoise. Concho belt. Multiple stones. Silver setting. Navajo. 3½" × 2¾". *JR's Southwestern Trading Post.*

Sleeping Beauty and Kingman turquoise. Concho belt. Turquoise and coral. Multiple stones, cluster. Silver setting. Navajo. 2¾". *JR's Southwestern Trading Post.*

Sleeping Beauty turquoise. Watch band. Multiple stones, inlay. Silver setting. Navajo. 6" × 1". *JR's Southwestern Trading Post.*

Sleeping Beauty turquoise. Rings. Turquoise, coral, shell, and black onyx. Multiple stones, inlay. Gold setting, left; Silver setting, right. Zuni.1"-1¼". *JR's Southwestern Trading Post.*

Sleeping Beauty turquoise. Rings. Turquoise, coral, shell, and black onyx. Multiple stones; left, chip inlay. Silver setting. Navajo. 1¼" × 1¼", 2" × 1". *JR's Southwestern Trading Post.*

Sleeping Beauty turquoise. Cuff bracelet. Multiple stones, cluster. Silver setting. Navajo, signed Verdy J. 3" × 1½". *JR's Southwestern Trading Post.*

Sleeping Beauty turquoise. Novelty necktie. Multiple stones. Silver setting. Navajo, signed Billy and Betty Betoney. *Courtesy of John Miller.*

Sleeping Beauty turquoise. Bolo tie. Sun God. Multiple stones. Silver setting. Navajo, signed J. Delgarito. *Courtesy of John Miller.*

Sleeping Beauty turquoise. Sculpture. Silver. Navajo, signed Wilford Begay. 9" height. *JR's Southwestern Trading Post.*

Sleeping Beauty turquoise. Sculpture. Silver. Navajo, signed Wilford Begay. 9" high. *JR's Southwestern Trading Post.*

Sleeping Beauty turquoise. Pendant necklace. Turquoise and coral. Design harkens to a Morrish horse bridal head-stahl motif; linking the history and evolution of silversmithing from pre-Columbian Morrish occupied Spain to the New World. Multiple stones. Silver setting. Zuni, signed Dickie and Darlene Charlie (D/D C). C. 1970–1980s. *Courtesy of John Miller.*

Sleeping Beauty turquoise. Squash Blossom necklace. Turquoise and coral. Reversible petite cut. Multiple stones. Silver setting. *Courtesy of John Miller.*

South American and Mexican (South America, Mexico) turquoise has been mined for centuries from a number of areas in the northern Mexican state of Sonoro, adjacent to the American Southwest, and in the western portion of South America. The material, found in shades of blue and green, is not easily distinguishable from North American turquoise, and is most often identified through provenance. It is an important historical source for the collector because of the early mining that was done, though little information is available on the individual mines or countries where the material has come to market.

Peruvian turquoise. Belt buckle. Silver setting. Signed John Renner. 6" × 4". *Courtesy of the artist.*

Peruvian turquoise. Concho belt. Multiple stones. Silver setting. Signed John Renner. 4" × 3½", buckle; 3¾" × 3", Concho. *Courtesy of the artist. JR's Southwestern Trading Post.*

Campito turquoise. Mexico. Bolo and buckle. Multiple stones. Silver setting. Navajo, signed Tommy Jackson. Bracelet. Silver setting. Navajo, signed Elle C. Jackson. Cabochons, Western Trading Post. *Photo courtesy of Western Trading Post.*

Stennich (Nevada) mine is one of the few that yield turquoise in predominantly green shades. A wide array of greens can be found in Stennich turquoise that make beautifully set pieces when cut and polished. It is a hard, high-grade material ideal for use alone or with other turquoises and gems.

Stennich turquoise. Belt buckle. Multiple stones, cluster. Silver setting. Navajo, signed CY. 1¾" × 1½". *JR's Southwestern Trading Post.*

Stennich turquoise. Belt buckle. Multiple stones, cluster. Silver setting. Navajo, signed CY. 2" × 1¾". *JR's Southwestern Trading Post.*

Stormy Mountain (Nevada) mine turquoise is distinctive for its dark blue stones. This material will often include a blotchy, black chert matrix resembling storm clouds, from which the mine got its name. Stormy Mountain turquoise can be confused with the similar-looking material mined from Blue Diamond.

Stormy Mountain turquoise. Bracelet. Silver setting. C. 1940s–1950s. *Courtesy of John Miller.*

Stormy Mountain turquoise. Pendant and necklace. Turquoise, coral, and black onyx. Multiple stones, inlay with beads. Silver setting. Navajo. 26". *JR's Southwestern Trading Post.*

Stormy Mountain turquoise. Pendant. Silver setting. Navajo. 6" × 1½". *JR's Southwestern Trading Post.*

Sunnyside (Nevada) mine is in northern Nevada near the town of Tuscarora in the Tuscarora mountain range. The Sunnyside was mined almost exclusively in the 1970s. The turquoise mine is no longer in operation as it has become part of a gold mining operation and a privately owned ranch. At the height of production, a significant quantity of material was shipped to Arizona and New Mexico and sold to Native American craftsmen. Sunnyside turquoise is most often dark blue and very hard, though beautiful green and blue/green colors can be found as well. With a spider web matrix of colors ranging from golden brown to black, these accent matrix markings are still sought after for use in custom jewelry.

Tiffany (New Mexico) mine is southeast of Santa Fe in the Cerrillos Turquoise District. The Tiffany mine has been worked for hundreds of years. The Anasazi Indians once worked the more than five separate mines within the sixty-acre area that comprises the Cerrillos mines. The Tiffany mine is named after Lewis Comfort Tiffany, who once owned and visited the mine. The turquoise was shipped by train to Albuquerque, then on to New York and crafted into exquisite jewelry. In 1892, George Kunz announced that certain colors of turquoise had come to be considered gem quality, namely, the Tiffany blue color. That year, seeking to mine the blue gem, James P. McNulty came to Cerrillos. McNulty took a job with the American Turquoise Company, which owned the claims to a number of mines. The ATC sold almost all its turquoise directly to Tiffany & Co., where well-known designer Pauling Farnham crafted more than $2 million worth into jewelry. Today, the Tiffany mine and five others in Cerrillos are owned by Doug Magnus, a Santa Fe jewelry designer. It is believed these mines have been played out for years.

Timberline (Nevada) mine is one of Nevada's smallest worked turquoise mines. The high-grade material from Timberline was limited in quantity and ranges from medium to dark blue in color, often with brown to black matrix. A rather difficult material to attain, it is a natural addition to a comprehensive collection.

Timberline turquoise. Assorted cut and polished stone. 14 grams. *Courtesy of John Renner.*

Timberline turquoise. Cuff bracelet. Multiple stone. Silver setting. Signed John Renner. 1¼" × 3". *JR's Southwestern Trading Post.*

Turquoise Mountain (Arizona) mine, near the famed Kingman mine, was closed in the late 1980s. The material from this mine is a lovely light to dark blue and can be found with or without spider webbing. The name birds-eye refers to turquoise from the mine that shows an area of blue circled by a darker blue matrix, resembling a bird's eye. A cut and polished piece from Turquoise Mountain is an outstanding addition to any collection.

Turquoise Mountain turquoise. Cut and polished stone. 4 grams. *Courtesy of John Renner.*

Turquoise Mountain turquoise. Assorted rough stones. 238 grams. *Courtesy of John Renner.*

Turquoise Mountain turquoise. Belt buckle. Multiple stones. Silver setting. Navajo. 5" × 4". *JR's Southwestern Trading Post.*

Turquoise Mountain turquoise. Cuff bracelet. Silver setting. Navajo. 3½" × 3". *JR's Southwestern Trading Post*

Turquoise Mountain turquoise. Pendant. Multiple stones. Silver setting. Navajo. 2½" × 1". *JR's Southwestern Trading Post.*

Turquoise Mountain turquoise. Bolo tie. Badger paw. Silver setting. Hopi, signed Victor Coochwytewa. Coochwytewa was a WWII veteran serving in Burma with the famed combat unit Merrill's Marauders. *Courtesy of John Miller.*

Tyrone (New Mexico) mine suffered a sad fate. Associated with the more valued copper mining operations southwest of Silver City, material mined from Tyrone is an intense high-grade blue stone. Unfortunately, no turquoise has been mined from Tyrone since the early 1980s, when miners began processing the copper ore through an acid wash rather than the more widely used crushing operation. The acid destroyed any turquoise that existed in the copper ore. A well-known mineral streak begins east of Silver City and wends its way through Arizona and into Mexico. This streak includes many outcrops and mines of various well-known Southwest minerals, including the turquoise of Morenci.

Tyrone turquoise. Belt buckle. Silver setting. Navajo, signed NB. 3¼" × 3". *JR's Southwestern Trading Post.*

Tyrone turquoise. Watch bracelet. Multiple stones, inlay. Silver setting. Navajo. 1" × 3". *JR's Southwestern Trading Post.*

Tyrone turquoise. Cuff bracelet. Silver setting. Navajo. 3" × 3". *JR's Southwestern Trading Post.*

Valley Blue (Nevada) was a small mine in the Lander County region between Austin and Battle Mountain. This mine was well-known and worked for a short time in the 1960s and 1970s. It produced a pleasant medium blue to dark blue turquoise with black matrix, and was most often found in nuggets from which semi-nugget cabochons were cut and polished.

Villa Grove (Colorado) mine, once known as the Hall mine, is northwest of the town of Villa Grove in the San Luis Valley. Turquoise from this mine is principally sky blue without any matrix. The mine was celebrated for its small, geologically perfect turquoise-colored nuggets. The turquoise from the Villa Grove can be a gorgeous robin's egg blue, so evenly colored that it appears to glow. While the nuggets were almost always the size of a small bean, occasionally dime-size nuggets were mined. Because of its uniform color, Villa Grove turquoise was cut and polished for setting in Zuni petite point and needlepoint jewelry, as well as Navajo cluster jewelry and Santo Domingo bead necklaces. A small amount of spider web matrix material was found at the mine but considered inconsequential, as is the rare occurrence of green turquoise. Randy Christensen purchased Villa Grove in 2000 and has used the turquoise in his own jewelry.

Valley Blue turquoise. Bracelet. "White" turquoise. Silver setting. Navajo, signed Toby Henderson. *Courtesy of John Miller.*

Related Minerals

Chalcosiderite (Southwest) is a cousin of turquoise that has found popularity among southwestern gem collectors. Sometimes referred to as white turquoise, Chalcosiderite does not meet the traditional mineralogy formula definition of true turquoise. Chalcosiderite is a rare copper phosphate mineral that consists of iron and aluminum, like turquoise, and contains more aluminum than iron.

Variscite (Utah) is a cousin of turquoise. It is a very hard stone and occasionally has a blue to green coloring, giving it the look of natural turquoise. While Utah variscite mines are pretty well depleted, for a number of years a fair amount of material was released into the marketplace. With its color resembling turquoise and degree of matrix and webbing, variscite became a favorite of jewelers and silversmiths. While it is difficult to find a good piece of variscite in Southwest jewelry today, it is an important adjunct mineral to add to a comprehensive collection, and looks beautiful cut, polished, and set in its own right.

Damele Variscite. Assorted cut and polished stones. 6 grams. *Courtesy of John Renner.*

Variscite. Necklace. Utah Broken Arrow variscite. Silver setting. C. 1980s. Navajo, signed Eddy Chaco. *Courtesy of John Miller.*

Variscite. Pendant. Silver setting. Navajo, signed E. 2½" × 1¾". *JR's Southwestern Trading Post.*

Variscite. Earrings. Multiple stones. Silver setting. Navajo. 1¼" × ½". *JR's Southwestern Trading Post.*

Variscite. Cuff bracelet. Multiple stones. Silver setting. Navajo. 1¾" × 3". *JR's Southwestern Trading Post.*

White Turquoise (Arizona, Nevada) is a controversial topic for turquoise purists. In fact, while some have called it a "variety" of turquoise, it is actually a rock consisting primarily of quartz, calcite, and alunite with only the slightest trace of turquoise. The fact is, mines in the United States have light to white turquoise material that will test positive as turquoise; however, it is rarely harder than chalk. Stabilizing it brings out the stone's color, and it is no longer white turquoise. Some US mines produce white turquoise that is hard enough for lapidary work, though this is rare, and stones that can be cut and polished are rarer still.

Reconstituted turquoise. Pendants. Multiple stones, cluster. Silver setting. Navajo. 1", 2" × 1½", × 3". *JR's Southwestern Trading Post.*

Reconstituted turquoise. Belt buckle. Multiple stones, cluster. Silver setting. Navajo. 4" × 3". *JR's Southwestern Trading Post.*

Reconstituted and Block Turquoise

Reconstituted turquoise. Necklace. Multiple stones, five-strand bead. Navajo. 24". *JR's Southwestern Trading Post.*

Reconstituted turquoise. Cuff bracelet. Turquoise and coral. Multiple stones, inlay. Silver setting. Navajo. 1½" × 3". *JR's Southwestern Trading Post.*

Reconstituted turquoise. Bolo tie. Multiple stones, cluster. Silver setting. Navajo. 3¼". *JR's Southwestern Trading Post.*

Reconstituted turquoise. Cuff bracelet. Turquoise, coral, shell, and black onyx. Multiple stones. Silver setting. Navajo. 2¼" × 3". *JR's Southwestern Trading Post.*

Reconstituted turquoise. Bolo tie. Multiple stones, cluster. Silver setting and tips. Navajo. 2½" × 2". *JR's Southwestern Trading Post.*

Block turquoise. Earrings. Multiple stones, inlay. Silver setting. Navajo. ¾" × 3/8". *JR's Southwestern Trading Post.*

Block turquoise. Earrings. Multiple stones. Navajo. 3¼" × 1½". *JR's Southwestern Trading Post.*

Reconstituted Kingman turquoise. Fetish necklace. Multiple stones. Navajo. 18". *JR's Southwestern Trading Post.*

Other Turquoise Mines

The following US turquoise mines are not described or illustrated in this chapter. Some are small, long ago closed, little known to collectors, or simply have little information. This is not intended to cast them aside; on the contrary, the turquoise collector has many avenues in which to track down raw material or cut and polished stones. Whether tracking them down online or by visiting the large gem and mineral shows, with perseverance you are bound to locate stones from many of these nearly forgotten locales.

Basalt (Nevada)
Battle Mountain (Nevada)
Bernadino (California)
Birdseye (Arizona)
Blue Boy (Nevada)
Blue Creek (Nevada)
Blue Diamond (Nevada)
Blue Eagle (Nevada)
Blue Ice (Nevada)
Blue Medicine (Nevada)
Blue Moon (Nevada)
Blue Thunder (Nevada)
Blue Warrior (Nevada)
Blue Wind (Nevada)
Bonnie Blue (California)
Broken Arrow (Nevada)
Broken Bow (Nevada)
Christie Blue (California)
Cochise (California)
Colorback I and II (Nevada)
Copper World (California)
Cortez (Nevada)
Creede District (Colorado)
Crow Springs (Nevada)
Cyprus Sierrita (Nevada)
Dry Creek (Nevada)
East Camp (California)
El Producto (California)
Elkhorn (Cripple Creek) (Colorado)
Florence (Cripple Creek) (Colorado)
Frog Skin (Nevada)
Gilbert (Nevada)
Gold Acres (Nevada)
Halley's Comet (Nevada)
Hidden Treasure (Cripple Creek) (Colorado)
Hidden Treasure (Nevada)
Himalaya (California)
Holycross District (Colorado)
Indian Mountain (Nevada)
Ivanhoe (Nevada)
Josie May (Leadville) (Colorado)
Last Chance (Nevada)
Leaning Shack (Nevada)
Lickskillet (Manassa) (Colorado)
Little Gem (California)
Llanada Copper (California)
Lucky Peak (Nevada)
Mastrada (Nevada)
McGinnis (Nevada)
Middle Claim (California)
Miss Moffet (Nevada)
Mistie Mountain (California)
Monte Cristo (Nevada)
Montezuma (Nevada)
Nevada Blue (Nevada)
New Monster (California)
Northern Lights (Nevada)
O'Haver (Cripple Creek) (Colorado)
Papoose (Nevada)
Pinto Valley (Arizona)
Pirate #3 (Nevada)
Pixie (Nevada)
Prince (Nevada)
Red Mountain (Nevada)
Renea Clue (California)
Roanoke (Cripple Creek) (Colorado)
Royal Web (Nevada)
Smith Black Matrix (Nevada)
Smoky Valley (Nevada)
Starquest Grove (California)
Stone Mountain (Nevada)
Sugarloaf District (Colorado)
Summitville District (Colorado)
Thunderbird (Nevada)
Tina Gem (Nevada)
Tortoise (Nevada)
Troy Springs (Nevada)
Turquoise Bonanza (Nevada)
Turquoise Chief (Leadville) (Colorado)
Valley Blue (Nevada)
Verde Blue (Nevada)
White Creek (Nevada)
White Owl (Nevada)
White Stallion (Nevada)
Windy Ridge (Nevada)
Zuni (Nevada)

Glossary

Agate. Fine-grained variety of chalcedony, often with colored bands or irregular splotches.

Aluminum. Silvery-white, metallic element, the most abundant in the Earth's crust though found only in combination, chiefly in bauxite.

Amazonite. Green variety of microcline, often used as a semiprecious stone.

Anasazi. Otherwise known as Pueblo. This is the name given by the Spanish to Native Americans who lived in adobe or stone houses in the Southwest US.

Apatite. Common complex mineral consisting of calcium fluoride phosphate or calcium chloride phosphate; a source of phosphorus.

Aztec. People of central Mexico whose civilization reached its height during the Spanish conquest in the early sixteenth century.

Bolo tie. Necktie of thin cord fastened in front with a decorative clasp.

Cabachon. Any precious stone cut and polished, but not faceted of convex hemispherical or oval form.

Chalcedony. Microcrystalline, translucent variety of quartz, often appearing as milky or grayish in color.

Chert. Heavily compacted rock consisting nearly entirely of microcrystalline quartz.

Chonchoidal. Having elevations or depressions shaped like the inside surface of a bivalve shell.

Chrysocolla. Hydrous copper silicate mineral; occurs in compact green or blue masses.

Coin silver. Metal containing ninety parts silver and ten parts other metal, and made of melted down pre-1900 Mexican and United States coins.

Color enhanced. Chemically altered to change the color of a stone.

Color treated. Often used term meaning color enhanced.

Concha. Hammered silver that often resembles a shell, flower, or sunburst; often having designs.

Copper. Metallic element having a distinctive reddish-brown color.

Coral. Hard, variously colored, calcareous skeleton secreted by certain marine polyps; variety of colors.

Cryptocrystalline. Having a microscopic crystalline structure.

Doublet. Composite, or assembled from two or more components.

Enhanced. To increase or improve in value, quality, desirability, or attractiveness.

Epoxy. A resin used to bond two materials.

Gemstone. Any precious or semiprecious stone that can be cut and polished for use as a gem.

Gold. Bright and precious yellow metallic element not subject to oxidation or corrosion.

Hohokam. An American Indian culture of the central and southern deserts of Arizona existing approximately 450–1450 C.E., roughly at the same time as the Anasazis to the north.

Inlay. To decorate an object with layers of materials set into the surface.

Insoluble. Incapable of being dissolved.

Intarsia. An art or technique of decorating a surface with inlaid patterns.

Iron oxide. Any one of various oxides of iron, such as ferric oxide or ferrous oxide.

Jade. Either of two minerals, jadeite or nephrite, usually green. Often used in jewelry or carved.

Jasper. Compact, opaque, cryptocrystalline variety of quartz, most often colored red.

Lapidary. The process of cutting and polishing gems and minerals.

Lapis lazuli. Deep blue mineral composed mainly of lazurite with smaller quantities of other minerals.

Matrix. The portion of a rock in which coarser crystals or rock fragments are surrounded.

Mimetite. A mineral that occurs in pale yellow or brownish hexagonal crystals.

Mine. Place in or on the Earth where minerals may be extracted.

Mohs scale. Scale of hardness used in mineralogy, from softest (1) to hardest (10).

Mother-of-pearl. Hard, iridescent substance that forms the inner layer of certain mollusk shells.

Nanometer. One billionth of a meter.

Navajo. Largest tribal group in the United States, located in Arizona and New Mexico.

Onyx. Variety of chalcedony having straight parallel bands of alternating colors; often referred to as "jet" when black.

Opaque. Blocking the passage of radiant energy and especially light.

Persia. Ancient empire extending from Egypt and the Aegean to India; conquered by Alexander the Great 334–331 B.C.E.

Petit point. A stone shaped to a fine point that is often larger than needle point, characterized by being round, oval, or having one rounded end.

Pyrite. Common brass-yellow iron disulfide mineral having a metallic luster often referred to as fool's gold.

Pyromorphite. Mineral; lead chlorophosphate that occurs in crystalline and massive forms, most often a green, yellow, or brown color.

Reconstituted. Pulverized pieces of stone that are then stabilized and hardened with resins to achieve a natural appearance. A man-made substitute for natural turquoise.

Refractive index. The ratio of the speed of radiation (as light) in one medium (as air, glass, or a vacuum) to that in another medium.

Silver. White metallic element.

Specific gravity. The ratio of the density of any substance to the density of some other substance taken as standard, water being the standard for liquids and solids, and hydrogen or air being the standard for gases.

Spectroscope. A device used to measure the properties of light.

Sphalerite. Common mineral, zinc sulfide, typically containing iron and cadmium, occurring in yellow, brown, or black crystals or cleavable masses.

Spider web. Name for pattern, often even, of matrix in turquoise stone.

Squash blossom. Design or configuration akin to the flower of the squash plant.

Stabilized. Infusing natural stone with a clear epoxy resin under pressure, which permanently hardens the rock and deepens the color.

Subvitreous. Having the appearance of minerals, somewhat vitreous.

Sutter's Mill. The location in California, near Sacramento, where gold was discovered in 1848, precipitating the gold rush of 1849.

Tiffany. Charles Lewis, 1812–1902, US jeweler; Louis Comfort, 1848–1933, US painter and decorator, especially of glass. Tiffany is also known for high-quality retail goods with distinctive blue packaging.

Translucent. Allowing light to pass through but diffusing it.

Transparent. Able to be seen through with clarity; so thin as to transmit light.

Treated. Subjected to a physical (or chemical) treatment or agent.

Turquoise. Hydrous phosphate of aluminum and copper used as a gem. It occurs rarely in crystal form, but is usually cryptocrystalline. Turquoise is opaque and often has a waxy luster; the color varies from pale blue to sky-blue, dark blue, green, white, and light brown.

Variscite. Secondary hydrated phosphate of aluminum mineral, occurring mainly as substantial bluish-green nodules

Vein. Area in host rock in which gem material is found; formed by seeping water in open area of rock.

Wampum beads. Cylindrical beads made from shells, pierced and strung.

Zachery (or Foutz) treatment. A proprietary process used to enhance turquoise.

Zinc. Bluish-white metallic element.

Zuni. Native North American Indians in western New Mexico inhabiting the largest Indian pueblos.

Bibliography

Andress, Donna. *Eldorado Canyon and Nelson, Nevada (Historical documents, reminiscences, commentary)* 1st edition. Compiled by Donna Andress. Nelson, Nevada: D. Andress, 1997.

Chalker, Kari. *Totems to Turquoise: Native North American Jewelry Arts of the Northwest and Southwest*. New York, N.Y.: Harry N. Abrams, 2004.

Dubin, Lois Sherr. *North American Indian Jewelry and Adornment*. New York: Harry N. Abrams, 1999.

Foxx, Jeffrey Jay and Carol Karasik. *Turquoise Trail: Native American Jewelry and Culture of the Southwest*. New York: Harry N. Abrams, 1993.

Holden, Martin. *The Encyclopedia of Gemstones and Minerals*. New York: Michael Friedman Publishing Group, Inc., 1991.

Johnsen, Ole. *Princeton Field Guide. Minerals of the World*. Princeton, NJ: Princeton University Press, 2002.

Kolbe, Regina. "Tribal Jewelry Hot Again with Western Look," *Antique Trader*. F+W Publications, August 10, 2005.

Kuhns, Elizabeth. "Wild About Turquoise," *Lapidary Journal*. Primedia Co., 2003.

Lowry, Joe Dan. *Turquoise Unearthed: An Illustrated Guide*. Santa Fe, NM: Rio Nuevo Publishers, 2002.

Mitchell, James R. *Gem Trails of Arizona*. Baldwin Park, CA: Gem Guides Book Co., 2001.

Pough, Frederick H. *A Field Guide to Rocks and Minerals*, 5th edition. New York: Houghton Mifflin, 1998.

Schiffer, Nancy. *Jewelry by Southwest American Indians: Evolving Designs*. Atglen, PA: Schiffer Publishing Ltd., 1997.

Schiffer, Nancy. *Turquoise Jewelry*, 2nd edition. Atglen, PA: Schiffer Publishing Ltd., 2004.

Simmons, Marc. *Turquoise and Six-Guns: The Story of Cerrillos, New Mexico*, 3rd edition. Santa Fe, NM: Sunstone Press, 1989.

Stacey, Joseph. *Arizona Highways*. Phoenix: Arizona Department of Transportation, January 1974.

Turnbaugh, William A. and Sarah Peabody Turnbaugh. *Indian Jewelry of the American Southwest*. Atglen, PA: Schiffer Publishing, 1997.

Vigil, Arnold (ed.), "The Allure of Turquoise," *New Mexico Magazine*; 2nd edition, 2005.

Zucker, Benjamin. *Gems and Jewels: A Connoisseur's Guide*. New York: Thames and Hudson Publishing Co., 1984.

www.bluearticles.com

www.encyclopedia.jrank.org

www.gemselect.com

www.geology.com

www.minerals.net

www.thesantafesite.com

www.turquoisecollector.com

www.totallywestern.com

Index

Numbers in italic denote photographs.

MARK P. BLOCK is the proprietor of Mark Block Originals + Fine Art and a long-time collector of gems and minerals, particularly varieties of quartz, calcite, fluorite, and turquoise. He is the author of *Contemporary Marbles and Related Art Glass* (Schiffer Publishing, 2001), and *The Encyclopedia of Modern Marbles, Spheres, and Orbs* (Schiffer Publishing, 2005), and is the coauthor, with Robert Block, of *Las Vegas Lights* (Schiffer Publishing, 2003). Block holds degrees in graphic design and political science, as well as a master's of business administration/marketing. He lives in Connecticut with his wife and daughter. Mark Block can be contacted at mark.block@hotmail.com or www.blockglass.net, or through Schiffer Publishing.